MW01630373

Sex, Gender, and Emerging Technology in Healthcare: Mitigating Bias and Fostering Equity

Heisook Lee · Nayoung Kim ·
Sabine Oertelt-Prigione
Editors

Sex, Gender, and Emerging Technology in Healthcare: Mitigating Bias and Fostering Equity

From Biology to Care: Sex and Gender Impacts on Health and Medicine

Springer

Editors
Heisook Lee
Korea Center for Gendered Innovations in Science and Technology Research (GISTeR)
Seoul, Korea (Republic of)

Nayoung Kim
Department of Internal Medicine
Seoul National University Bundang Hospital
Seongnam, Korea (Republic of)

Sabine Oertelt-Prigione
Department of Primary and Community Care
Radboud University Medical Center
Nijmegen, The Netherlands

ISBN 978-981-95-2069-5 ISBN 978-981-95-2070-1 (eBook)
https://doi.org/10.1007/978-981-95-2070-1

This Springer imprint is published by the registered company Springer Nature Singapore Pte Ltd.
The registered company address is: 152 Beach Road, #21-01/04 Gateway East, Singapore 189721, Singapore

Preface

The impact of sex and gender on biological phenomena and health is being robustly documented through a growing body of evidence. Spearheaded by a number of active pioneers in the field and supported by regulatory developments in the last decade, such as the mandates by funding agencies and scientific journals to include a focus on sex and gender in grants and publications, knowledge production has increased substantially. The impact of sex differences on the development, progression and therapeutic response in fields such as cardiology, oncology, and immunology is now widely accepted. Other disciplines are catching up, as exciting developments in the field of neurology, gastroenterology, endocrinology, metabolomics, and pharmacology are showing us. A growing focus on the development of methods to measure gender in biomedical research is also proving essential in uncovering its impact on medical access and communication, as well as on patients' quality of life and their perception of care. Taken together, these developments can truly enhance our progress toward a more holistic, individualized, and patient-centered care in the future.

Nevertheless, we also have to acknowledge that Sex- and Gender-Specific Medicine (SGSM) is a developing field where much remains to be uncovered and explained. Novel developments are visible in the traditional sectors of medicine, such as clinical and laboratory research, but they also evolve in new directions given the exponential adoption of digital technologies and generative algorithms. These dynamics were at the core of the organization of the 2023 Conference BrainLink X-Lab Day Exploring Life Phenomena: Integrating Clinical Science and Artificial Intelligence for Inclusive Research. Held in Pyeongchang and Seoul from November 13 to 15, 2023, this pioneering interdisciplinary event was initiated by the Korea Center for Gendered Innovations in Science and Technology Research (GISTeR) and generously funded by the Korea Federation of Science and Technology Societies of Republic of Korea. The first of this kind, the interdisciplinary conference brought together scientists working on different, yet complementary aspects in the SGSM field. The participants included specialists in gastroenterology, public health experts, neuroscientists, AI researchers, and methodological and implementation specialists. State-of-the-art knowledge and new developments were shared to pave a new research agenda that can shape the developments in the SGSM field in Korea, Asia, Europe, and the USA as a whole.

The current publication assembles the different methodological and topical approaches in one structured compendium, drawing from the presentations, and discussions held at the BrainLink conference, themed "Exploring Life Phenomena: Integrating Clinical Science and Artificial Intelligence for Inclusive Research." Spanning from the history, operationalization and relevance of SGSM as a cross-sectional approach, it then highlights disease-specific applications in the fields of gastroenterology, oncology, and neuroscience. Then a life course approach that centers on the impact of the environment on sex and gender as well as the impact of sex and gender on pharmacology are discussed. Subsequently, novel developments in the field of machine learning and AI are described with a special focus on ethical approaches and inclusivity. Overall, a comprehensive picture of the multifaceted approaches emerges and highlights how diverse methodologies and targets in the field are not only a fact but also a resource that generates vibrant discussions and connects field-specific priorities with a cross-cutting sex- and gender-sensitive approach.

In the first part *Introduction to Sex and Gender in Biomedical Science*, the author emphasizes that a lack of appropriate representation and robust analysis leads to a lack of inclusive clinical decision-making tools and limits the ability to predict therapeutic efficacy for all patients. She suggests that sex and gender, in intersection with other social and demographic factors, can impact health and disease in many distinct ways. In the second part *Disease-Specific Sex Differences*, authors explore sex and gender differences in gastric cancer and hepatocellular carcinoma (HCC). The author found that estrogen receptor β (ERβ) provides a protective effect against gastric cancer, particularly in the intestinal and Epstein–Barr virus (EBV)-associated types. She also noted that immune checkpoint inhibitors (ICIs) targeting PD-1 and PD-L1 were more effective for men, suggesting sex-based differences in immune responses. In HCC, the authors identified a key molecular link between Gα12 signaling and estrogen pathways, specifically involving estrogen receptor α (ERα). This interaction appears to be the driving force behind the sexual dimorphism observed in HCC. In the third part *Sex Differences in Neuroscience and Brain Health*, the authors explain why sex differences in neuroscience hold critical implications for understanding mental health disparities and tailoring interventions. For instance, neuroimaging research on sex differences has highlighted sex-specific vulnerabilities: for example, men are more frequently affected by autism spectrum disorder, Parkinson's disease, traumatic brain injury, and strokes, while women are more prone to posttraumatic stress disorder, mood disorders (bipolar/depression), Alzheimer's disease, and multiple sclerosis. The authors provide excellent examples of sex- and gender-related diseases in neuroscience. In the fourth part *Environmental and Lifespan Considerations*, the authors introduce that the effect of prenatal stress on physiology and behavior has been mainly studied in male offspring in rodent models, but gender differences have been investigated through human studies. Thus, they discuss the various effects of maternal stress, and differences by gender. The authors also demonstrate that the environmental chemicals affect women's health from mid- to late-life based on the Study of Women's Health Across the Nation

Multi-Pollutant Study (SWAN-MPS). Additionally, the importance of improving drug safety, advancing personalized medicine, and achieving more equitable healthcare outcomes is emphasized. The final part, *Innovative Approaches and Emerging Technologies*, introduces machine learning, particularly in its more practical forms, as a tool for offering more standardized, consistent, objective, inclusive, equitable, and efficient assessments and decision-making. Moreover, the authors argue that AI can be transformed from a bias amplifier into a powerful instrument for achieving genuinely equitable, transparent, and patient-centered healthcare by using an "equity-by-design" approach that prioritizes sex- and gender-sensitive methodologies and intersectional analysis.

Although broad in their spectrum, the different chapters highlight some common features that are also a to-do-list for the future. In all of the different fields, lack of sex- and gender-specific data is evident, most frequently to the detriment of women and female patients. Historical shortcomings in the consideration and enrollment of women and female participants in research have led to a persistent gender data gap, which affects much of the clinical choices and digital applications developed today. Inadequate representation and a lack of rigorous analysis result in a deficiency of inclusive clinical decision-support systems, thereby constraining the capacity to forecast therapeutic efficacy across diverse patient populations. The same lack of representation also hinders the development of ethical and inclusive AI solutions, even when awareness of historical biases is present.

The different chapters emphasize how new knowledge is being produced to close these historical gaps, but also how focus, dedication, and funding is needed to make this process a reality. The gender data gap will not close itself—awareness, dedication, and support are needed to make healthcare inclusive for all. Especially in times when the role of biomedical research is being undermined and the role of scientific progress questioned, we as scientists have an ethical and moral duty to support a better future for our societies, by providing methods, evidence, and policy frameworks that can improve the health for all. We hope this book contributes to these goals.

Seoul, Korea (Republic of) — Heisook Lee
Seongnam, Korea (Republic of) — Nayoung Kim
Nijmegen, The Netherlands — Sabine Oertelt-Prigione

Acknowledgments

This book is the culmination of a collaborative effort, and we extend our deepest gratitude to everyone who contributed to its publication.

The foundation for this book was laid during the "BrainLink X-Lab Day: Exploring Life Phenomena: Integrating Clinical Science and Artificial Intelligence for Inclusive Research" conference. This event, generously sponsored by the Korea Federation of Science and Technology Societies (KOFST) in 2023, brought together prominent scholars, young scientists, and students to discuss a crucial issue: the need to incorporate diverse factors, such as sex and gender, to advance research and promote equitable healthcare. This vital initiative was spearheaded by the Korea Center for Gendered Innovations in Science and Technology Research (GISTeR).

Held from November 13 to 15, 2023, in Pyeongchang and Seoul, the conference served as an invaluable platform for sharing insights and inspiration. The participants collectively decided to compile the presentations and discussions into a book to widely disseminate the importance and challenges of incorporating sex- and gender-based research. We are especially grateful for the dedicated efforts of Dr. Heajin Kim and Dr. Chin-Hee Song, who played pivotal roles in this process.

Following the conference, we invited additional researchers who contributed outstanding work to join as co-authors. This collaboration resulted in the completion of ten exceptional chapters. We thank all the authors for their hard work and dedication in elevating the quality of this book.

We also wish to express our profound appreciation to the outstanding editors, Prof. Nayoung Kim and Prof. Sabine Oertelt-Prigione, and to Ms. Jihyun Park and Dr. Heajin Kim for their invaluable administrative and editorial support.

Finally, we are immensely grateful to the Korea Center for Gendered Innovations in Science and Technology Research (GISTeR) for their unwavering administrative and financial support, without which this publication would not have been possible.

On behalf of the editors.

Heisook Lee

Contents

Part I
Introduction to Sex and Gender in Biomedical Science

The Role of Sex and Gender for Health and Disease

Sabine Oertelt-Prigione

Abstract Sex and gender are progressively recognized as important determinants of health in the biomedical field. In intersection with other social and demographic factors, sex and gender, can impact health and disease in many distinct ways. They affect the emergence of disease, by way of genetics, hormonal influences and exposure to risk factors, as well as the progression of disease and its response to therapy. In addition, they can modulate access to health care and chances of recovery. In the following chapter, the operationalization of sex and gender will be described and several examples of their impact on disease development will be provided.

Keywords Sex · Gender · Data · Operationalization · Translation · Precision

1 Brief History of Sex and Gender-Sensitive Medicine (SGSM)

Gendered stereotypes have characterized medicine throughout its history. The perception of the female body as “atypical” or “other” in comparison to a standard male body has shaped medical thinking since ancient times. In Hippocratic times (fifth century BC) women’s bodies were described as “porous” and prone to the accumulation of fluids as well as characterized by a wandering uterus—*hystera*—which could move throughout the body leading to symptoms of hysteria. The idea that female reproductive organs were somewhat connected to the emergence of disease, especially the “aspecific” and psychosomatic ones, was carried into Medieval times where the womb was still described as “the origin of all diseases” [1].

S. Oertelt-Prigione (✉)
Gender Unit, Department of Primary and Community Care, Radboud University Medical Center, Nijmegen, Netherlands
e-mail: sabine.oertelt-prigione@radboudumc.nl

Sex- and Gender-Sensitive Medicine Unit, Medical Faculty OWL, University of Bielefeld, Bielefeld, Germany

H. Lee et al. (eds.), *Sex, Gender, and Emerging Technology in Healthcare: Mitigating Bias and Fostering Equity*, https://doi.org/10.1007/978-981-95-2070-1_1

These historical assumptions are still (partially) reproduced in modern times and gave rise to the women's health movement in the 1960 and 1970 and followingly to sex- and gender-sensitive medicine (SGSM). During the second wave of feminism in the 1960 and 70, the focus in the health field was mostly placed on sexual and reproductive rights, among others on the de-medicalization of childbirth, access to contraception and safe abortion [2]. A milestone in this development in the United States was the publication in 1971 of "Our Bodies, Ourselves", a manifesto from the Boston Women's Health Book Collective (http://ourbodiesourselves.org). In parallel, institutional actions were established in the USA, Canada and Europe to formalize the inclusion of sex and gender in both research practice and enshrine principles of equity into institutions. Examples of these measures are the European directive to eliminate discrimination on the grounds of sex, the Equal Opportunity in Science and Engineering Act in the US and the NIH Health Revitalization Act in 1993, which defined the legal grounds for the inclusion of women and minorities in clinical trials. The NIH established the Office of Research on Women's Health (ORWH), the first national agency focusing on women's health issues, and the European Union founded the European Institute of gender Equality (EIGE), which mainly focuses on gender mainstreaming but also targets several health-related themes [3].

During these decades a shift from a strict focus on women's health to a focus on sex and gender and their impact on health and disease as a whole emerged. As described by prof. Marianne Legato, one of the founders of gender-sensitive medicine in the United States, the movement shifted away from a "bikini medicine" approach—only considering differences for organs and body parts covered by a bikini—to a more holistic investigation of sex- and gender-related differences [4, 5]. An essential document in this shift was the 2001 landmark report by the US Institute of Medicine "Exploring the biological contributions to human health: does sex matter?". The report laid out in detail why consideration of sex differences is essential in biomedical research by emphasizing the role of genetics and hormones even at the cellular level.

The progressive focus on sex and gender, rather than women's health, also translated into research policy. For example, in 2004, the International Conference on Harmonization of Technical Requirements for Registration of Pharmaceuticals for Human Use (ICH) published the first regulation calling for the inclusion of women and men in clinical trials. The document was revised in 2009 to require also a differentiation of sex and gender. The Council of Europe declared the need for gender mainstreaming in health policy in 2008 and emphasized the importance of gender-sensitive medicine, while Canada started implementing Sex and Gender-based Analysis (SGBA) in 2009 to review health policy laws and regulations. The European Commission also developed a similar toolkit to support researchers and reviewers in their activities within the framework programs (https://op.europa.eu/en/publication-detail/-/publication/c17a4eba-49ab-40f1-bb7b-bb6faaf8dec8). In fact, since 2002 with the beginning of the 6th Research Framework Program (FP6), the European Commission has requested that researchers focus on the three gender dimensions within their proposals: 1. Gender equality in scientific careers, 2. Gender balance in decision making and 3. Integration of gender dimension in to the content of research and innovation. These recommendations were further refined in Horizon 2020 (FP8), with the definition of recommended targets and integrated with a request to combine every application with a gender equality plan (GEP) in 2022 under the Horizon Europe framework program. In 2011, the United Nations (UN) published their Resolution on Gender Analysis in Scientific Education, Research, and Development. It resolved that "gender-based analysis and gender impact assessments" should be integrated into science and technology research and development, and that "a gender perspective" should be incorporated into all science and technology curricula.

Since 2014 the US-American NIH also requested that applicants include female and male cells and animals in their research with the final goal to balance sex-based inclusion in basic biomedical research. Since 2014 the Canadian Institute of Health Research (CIHR) also requires all grant applicants to respond to mandatory questions about sex and gender in research [3].

Even editors are progressively focusing on the incorporation of sex and gender into scientific publications. The International Committee of Medical Journal Editors (ICMJE) has developed a constantly updated framework on sex and gender-sensitive reporting. The guideline recommends using the correct terms for sex and gender, reporting sex and/or gender of study participants, describing sex of animals or cells and explaining the methods to determine sex and gender. In 2012 the European Association of Science Editors (EASE) established a Gender Policy Committee and tasked it with the development of a set of guidelines for reporting of sex and gender research "Sex and Gender Equity in Research" (SAGER). The SAGER guideline, was published in 2016, and functions as a standard for both authors and reviewers [6]. It is currently recommended by a number of science publishers as a standard for reporting of sex and gender in biomedical research.

2 Operationalization of Sex and Gender for Biomedical Research

The growing attention to sex and gender from both funding agencies and publishers demands robust and reproducible methods. In order to produce high quality research in the field of SGSM the object of the investigation needs to be clearly stated, the analytical method described in detail and the data reported in a FAIR (findable, accessible, interoperable and reusable) manner. This will eventually increase the scientific value of the published research and its acceptance within the wider community.

An important first step is the clear definition of the object of study. This often means asking the question: are sex or gender differences being investigated? In the biomedical literature the two are still often used interchangeably leading to methodological confusion [7, 8]. In biomedical research, ***sex*** is defined by biological and physiological characteristics, such as chromosomes, sex organs, and endogenous hormonal profiles. These characteristics can be summarized with the acronym "3G", encompassing genes, gonads (hormones) and genitalia [9]. Using this framework sex is operationalized as female, male and intersex. Nevertheless, the varying nature of sex-related variables has to be considered. For example, the karyotype of circulating blood cells might change with aging leading to a higher percentage of sex chromosome monosomy [10, 11]. Hormone levels change during the day, during the month and with aging [12]. Last, anatomy is not always unequivocal or binary, but characterized by variations in the general population [13, 14]. To perform state of the art sex-specific analysis these variations need to be considered and addressed by providing details about the operationalization of sex. ***Gender*** is a socially constructed category that changes in time and space and is performed in interaction with others. Historically, gender has been primarily studied in the social sciences leading to its operationalization and measurement in forms uncommon for biomedical research. Operationalizing gender is a challenge for medicine, since the construct is complex and multifaceted, which means that it cannot be adequately captured by a simple binary question. The commonly used approach to circumvent this challenge is the dissection of the gender variable into several different gender dimensions, which can be measured singularly or in combination [15, 16]. These dimensions are:

- Gender identity—defines the subjectively perceived identity of an individual, such as being a woman, a man, a non-binary person etc. Identity is not defined by sex assigned at birth (SAAB), but often matches it.
- Gender expression—describes the matching of one´s presentation as an individual with the culturally-defined stereotypes associated to a specific gender
- Gender roles and/or norms—represent socially constructed expectations and behaviors attributed to a specific gender identity and/or expression
- Gender relations—explain how gender impacts the interaction between individuals in the private, professional and public space mostly through power differentials
- Institutionalized Gender—depicts the broader impact of gender on organizations, societal structures, legislation and access to resources

These constructs can be measured in isolation, in combination or in the form of "gender indices", which combine items pertaining to several of the dimensions within one single questionnaire. The approach to the development of these questionnaires can be different, leading to questionnaires falling into two main categories: data-driven or theory-driven. The first ones employ statistical approaches—mostly principal component analyses (PCAs)—to define which variables in large datasets associate with SAAB, but are unrelated to SAAB. These are then compounded as gender proxies. The second approach is based on pre-existing work defining activities, attitudes and personality traits commonly associated with a specific gender. An example of the first type is the work by Ballering and colleagues that employed a 13,000 participant population to identify potential differences in common somatic symptom associations using either sex as a predictor or the gender index [17]. An example of the second is the Gender as a Sociocultural Variable (GASV) index developed by Nielsen and colleagues [18]. The index is based on three categories: gender norms, gender-related traits, and gender relations, but explicitly refrains from labeling specific characteristics as "masculine" or "feminine". A last example combing both approaches is the gender index developed by Pelletier and colleagues in the context of the GENESIS-PRAXY study [19, 20]. The authors defined a priori a number of gender-related psychosocial variables and then proceeded to a PCA to identify which of these were indeed associated with SAAB.

3 Impact of Sex and Gender in Practice

3.1 Clinical Implications

Sex and gender can impact the expression of symptoms, accuracy to diagnostic tools and response to pharmacotherapy.

Differences in symptom expression and disease presentation can originate from underlying biological differences affecting disease development or from gender differences in perception and interaction with the healthcare system [21]. Probably the most well-known example of sex/gender-related differences in symptom presentation is cardiac ischemia [22]. Historically, cardiac ischemia and heart attacks have been associated with chest pain irradiated to the arm and jaw. While this symptom combination is very common with cardiac ischemia, younger patients, especially women, might not display these types of symptoms [23]. In the younger female patient group the absence of symptoms or the presence of symptoms such as dizziness, nausea, back pain and dyspnea might replace the expected chest pain and lead to potentially delayed diagnoses. Differences in symptoms have been reported in many diseases. Typical asthmatic wheezing appears much more common in young male patients, while young female patients might only present with dry nocturnal cough [24]. Male patients with Parkinson´s disease appear more frequently affected by freezing of gait and rigidity in the early phases of disease, while female patients are more likely to present with evident tremor [25]. In the field of autoimmune disease, female patients with the autoimmune disease lupus erythematosus are likely to report malar rash and Raynaud´s phenomenon, which are rarely reported in male patients. This is of significant importance given that visible symptoms can prompt timely investigation of the disease. Indeed, in male patients lupus erythematosus diagnosis is often delayed and frequently confirmed once the patient already display internal organ damage [26, 27].

Diagnostic tools can also display differences in sensitivity and specificity. When testing patients for the presence of pre-diabetes, fasting blood sugar appears as a very specific test for male patients, while female patients might need an oral glucose tolerance test to achieve the same specificity [28]. Most likely due to more efficient, possibly hormonally-driven compensatory mechanisms in the early phases of disease in female patients, blood glucose is less predictive of pre-diabetes compared to a functional test. Fecal occult blood screening (FOBS) tools also display a similar pattern with different sensitivities for guaiac-based tests in female and male users due to physiological differences [29]. Guaiac-based tests react with hemoglobin in fecal matter, but do not identify hemoglobin once it is oxidized. Oxidization depends on permanence time in the colon, i.e. the longer the hemoglobin from blood cells derived from cancerous tissue stays in the colon lumen, the more likely it is to be oxidized. Distance of the tumor from the anus and colon transit velocity can both affect permanence time. Female patients are more likely to experience right colon cancer compared to male patients [30] and colon transit times are slower on average in female patients due to—among other factors—the myorelaxant function of progesterone. These factors combined can lead to less sensitivity of guaiac-based FOBS in female users.

Differences in sensitivity have also been reported e.g. for high-sensitivity troponine to detect cardiovascular events. High-sensitivity troponine performs significantly better than conventional troponine measurements in female patients, markedly increasing detection rates. The same pattern could not be identified in male patients [31].

In addition to differences in symptoms and diagnosis, differences in access to therapy have also been documented in many medical disciplines. These differences appear to be a combination of older age at presentation for female patients leading to more conservative approaches, different patient preferences or assumptions thereof, different expectations from healthcare providers and systemic bias in allocation of therapeutic options [32]. In many specialties, patients and providers can make a choice between conservative therapeutic approaches, such as pharmacotherapy or radiotherapy, or more invasive ones, such as surgery. Observation and the choice of delayed intervention is a third option. Less invasive therapeutic options for female patients have been reported in many fields, in cardiology, oncology, neurology and in the transplantation field to name a few [33, 34]. Therapeutic choices can sometimes appear unjustified by clinical differences between patients and hence, be attributed to other reasons ranging from different patient preferences to inequity in treatment allocation.

Different responses to pharmacotherapy have also been highlighted in clinical practice as well as in clinical trials and pharmacovigilance efforts and will be detailed in the next section.

3.2 Clinical Trials

Clinical trials executed today are generally designed to include to at least female and male participant, although the inclusion of female participants varies by discipline and is still insufficient in many fields. In fact, a recent analysis of more than 1,400 manuscripts published between 2015 and 2019 in three highly reputable medical journals demonstrated how the inclusion of female participants still lagged behind the one of male participants (56 vs. 44%) [35]. Even if average percentages of inclusion are approaching 50% in some disciplines, taking the sex-specific clinical impact and experienced disability into account demonstrates persisting inequalities especially in highly prevalent conditions [36]. Dekker and colleagues recently demonstrated how female participants were well represented in marketing authorization application dossiers submitted to the European Medicines Agency (EMA), yet their numbers in phase III trials oftentimes did not match disease prevalence rates in the general population and representation varied depending on the conditions studied [37]. More adverse side effects were reported in the female patients in the investigated treatment groups, however, this trend also persisted in some of the placebo groups. Importantly, the underrepresentation of female participants does not appear justified by more reluctance to participate in clinical trials. In fact, Gruca and colleagues not only reported no gender differences in the willingness to participate in clinical trials, but identified more favorable attitudes towards clinical trials in women compared to men [38].

Historically, however, the situation has been different. In fact, in 1977, after the thalidomide tragedy, the FDA ruled for the exclusion of females of childbearing potential from clinical trials. The risk of potential unknown teratogenic effects was deemed so important to neglect the health of potential future consumers of the tested pharmaceuticals—mandatory contraceptive use during the trial was not perceived as a potential solution at the time [39]. The consequences of this ban of women from clinical trials can be seen until today. Only in 1993 with the Revitalization Act did the FDA introduce a mandate to include women and minorities rectifying the prior decision. Nevertheless, trials conducted in those 15 years without the inclusion of female participants would not be repeated and information about potential sex-dependent differences in efficacy and potential side effects could only be detected with pharmacovigilance efforts. It is important to know that among the ten drugs retracted from the US market by the FDA between 1997 and 2000, eight demonstrated a significantly higher risk of potential side effects in female users [40].

It is important to emphasize that the participation of women in clinical trials has steadily increased since the 1990s, however, participation does not necessarily translate into consideration upon analysis. In fact, although many studies aim at a 50% inclusion rate, analyses oftentimes are not disaggregated by sex of the participants limiting the identification of potential differences in efficacy or side effects of the medication tested. These trends have not changed in recent years and even during the COVID-19 pandemic, when potential sex/gender differences in the virus infectiousness and morbidity were widely discussed in the media [41].

3.3 Transition to Precision Medicine

Precision medicine has been primarily identified as an opportunity to tailor treatment to individual patient characteristics, mostly defined by the genetic makeup of the patient themselves or e.g. the genetic signature of a tumor [42]. Measurable biological characteristics are employed as determinants, while social and cultural influences are mostly ignored as potential individualization characteristics. The rise of companion diagnostics and next generation sequencing as precision tools underscores these approaches. Precision medicine has historically been defined by the

P1–P4 model, including prediction, prevention, personalization, and participation [43]. Prediction is intended as the ability to identify potential diseases early and maximize the ability to slow their emergence and progression, as defined with the next step, prevention. Personalization represents the core concept of individualized diagnosis and treatment for each patient based on their unique characteristics and participation should underscore the need for active involvement of the patient in their therapeutic and care decisions. The approach is being increasingly applied and described as an ideal to strive towards for the medical care of the future. While promising in its overall focus on the patient's individuality and in its recognition of the limitations of highly controlled randomized clinical trials in representing individual patient characteristics, it falls nonetheless short in capturing the social nature of human beings. Patients live in a real-world setting where their opportunities, risks and choices are characterized by many interconnected factors. These factors are partially biological, but just as much of social nature. Characteristics such as age, social status, education, migration background, religion, professional choices, sexual orientation impact personalization as much as genetics. This complexity will have to be adequately represented in the precision medicine of the future [44–47].

Attention to sex and gender could be a first step in this direction. In fact, sex and gender are a perfect representation of two intertwined variables that cover the biological and social spectrum. Including a focus on sex and gender in precision medicine efforts could thus become a best practice example of the advantages of such a combined inclusion. As detailed in a recent publication, a true precision approach in current times cannot be disengaged from the community of practice and from the population as whole. We propose an expanded precision medicine model, named $P3^2$, which also comes to include the domains P5–P6 (psychosocial and percipient) and P7–P9 (population-informed, partnered with community and public-engaging) [48]. In this expanded concept *psychosocial* encompasses the multiplicity of factors influencing disease emergence and therapeutic choices, while *percipient* highlights the importance of precision in terminology and communication. *Population-informed* addresses the growing opportunities and challenges connected to the availability of large and diverse datasets. *Partnered with the community* highlights how precision approaches need to be embedded in the lived realities of participants and patients. Finally, *public-engaging* emphasizes the importance of interaction and feedback from the communities of practice in order to deliver a precision medicine that is in line with their demands. It also includes a focus on the growing trends of citizen science and self-monitoring both of which contribute new perspectives and rich data sets. Sex and gender should be considered at all levels of this $P3^2$ approach to allow for future-oriented holistic care for individual patients.

4 Conclusion

Sex and gender have emerged as factors that can significantly impact health and disease. The growing body of knowledge underscoring their relevance is increasing the overall openness of clinicians and researchers to incorporate this knowledge into their practice. However, the structural implementation of sex, gender and intersectional aspects is still far from being achieved and much effort is needed to further their implementation. In a biomedical system increasingly focused on personalized and individualized healthcare, these variables can no longer be neglected and should be structurally integrated in a contextualized manner to support excellent research and the best possible healthcare for all patients.

References

1. Cleghorn E. Unwell women: misdiagnosis and myth in a man-made world. New York: Dutton, Penguin Random House LLC; 2021. p. 386.
2. Schiebinger LL. Has feminism changed science? Cambridge, Mass., London: Harvard University Press; 1999.
3. Becher E, Oertelt-Prigione S. History and development of sex and gender sensitive medicine (SGSM). Int Rev Neurobiol. 2022;164:1–25.
4. Legato MJ. Beyond women's health the new discipline of gender-specific medicine. Med Clin North Am. 2003;87(917–37):vii.
5. Legato MJ. Gender-specific medicine: the view from Salzburg. Gend Med. 2004;1:61–3.
6. Heidari S, Babor TF, De Castro P, Tort S, Curno M. Sex and gender equity in research: rationale for the SAGER guidelines and recommended use. Res Integr Peer Rev. 2016;1:2.
7. Nieuwenhoven L, Klinge I. Scientific excellence in applying sex- and gender-sensitive methods in biomedical and health research. J Womens Health (Larchmt). 2010;19:313–21.
8. van den Hurk L, Hiltner S, Oertelt-Prigione S. Operationalization and reporting practices in manuscripts addressing gender differences in biomedical research: a cross-sectional bibliographical study. Int J Environ Res Public Health. 2022;19:14299.
9. Joel D. Genetic-gonadal-genitals sex (3G-sex) and the misconception of brain and gender, or, why 3G-males and 3G-females have intersex brain and intersex gender. Biol Sex Differ. 2012;3:27.
10. Guttenbach M, Koschorz B, Bernthaler U, Grimm T, Schmid M. Sex chromosome loss and aging: in situ hybridization studies on human interphase nuclei. Am J Hum Genet. 1995;57:1143–50.
11. Machiela MJ, Zhou W, Karlins E, Sampson JN, Freedman ND, Yang Q, et al. Female chromosome X mosaicism is age-related and preferentially affects the inactivated X chromosome. Nat Commun. 2016;7:11843.
12. Bailey M, Silver R. Sex differences in circadian timing systems: implications for disease. Front Neuroendocrinol. 2014;35:111–39.
13. Fausto-Sterling A. The five sexes, revisited. Sciences (New York). 2000;40:18–23.
14. Fausto-Sterling A. Intersex: concept of multiple sexes is not new. Nature. 2015;519:291.
15. Oertelt-Prigione S. The operationalization of gender in medicine. In: Legato M, editor. Principles of gender-specific medicine, vol. 2, 4th ed. New York: Elsevier; 2023. p. 503–12.
16. Tannenbaum C, Greaves L, Graham ID. Why sex and gender matter in implementation research. BMC Med Res Methodol. 2016;16:145.

17. Ballering AV, Bonvanie IJ, Olde Hartman TC, Monden R, Rosmalen JGM. Gender and sex independently associate with common somatic symptoms and lifetime prevalence of chronic disease. Soc Sci Med. 2020;253: 112968.
18. Nielsen MW, Stefanick ML, Peragine D, Neilands TB, Ioannidis JPA, Pilote L, et al. Gender-related variables for health research. Biol Sex Differ. 2021;12:23.
19. Pelletier R, Ditto B, Pilote L. A composite measure of gender and its association with risk factors in patients with premature acute coronary syndrome. Psychosom Med. 2015;77:517–26.
20. Pelletier R, Khan NA, Cox J, Daskalopoulou SS, Eisenberg MJ, Bacon SL, et al. Sex versus gender-related characteristics: which predicts outcome after acute coronary syndrome in the young? J Am Coll Cardiol. 2016;67:127–35.
21. Oertelt-Prigione S, Regitz-Zagrosek V. Sex and gender aspects in clinical medicine. London: Springer; 2012.
22. Regitz-Zagrosek V, Oertelt-Prigione S, Prescott E, Franconi F, Gerdts E, Foryst-Ludwig A, et al. Gender in cardiovascular diseases: impact on clinical manifestations, management, and outcomes. Eur Heart J. 2016;37:24–34.
23. Canto JG, Rogers WJ, Goldberg RJ, Peterson ED, Wenger NK, Vaccarino V, et al. Association of age and sex with myocardial infarction symptom presentation and in-hospital mortality. JAMA. 2012;307:813–22.
24. Almqvist C, Worm M, Leynaert B, working group of GALENWPG. Impact of gender on asthma in childhood and adolescence: a GA2LEN review. Allergy. 2008;63:47–57.
25. Gottgens I, van Halteren AD, de Vries NM, Meinders MJ, Ben-Shlomo Y, Bloem BR, et al. The impact of sex and gender on the multidisciplinary management of care for persons with Parkinson's disease. Front Neurol. 2020;11: 576121.
26. Jolly M, Sequeira W, Block JA, Toloza S, Bertoli A, Blazevic I, et al. Sex differences in quality of life in patients with systemic lupus erythematosus. Arthritis Care Res (Hoboken). 2019;71:1647–52.
27. Fish EN. The X-files in immunity: sex-based differences predispose immune responses. Nat Rev Immunol. 2008;8:737–44.
28. Kautzky-Willer A, Harreiter J, Pacini G. Sex and gender differences in risk, pathophysiology and complications of type 2 diabetes mellitus. Endocr Rev. 2016;37:278–316.
29. Brenner H, Haug U, Hundt S. Sex differences in performance of fecal occult blood testing. Am J Gastroenterol. 2010;105:2457–64.
30. Lee MS, Menter DG, Kopetz S. Right versus left colon cancer biology: Integrating the consensus molecular subtypes. J Natl Compr Canc Netw. 2017;15:411–9.
31. Shah AS, Griffiths M, Lee KK, McAllister DA, Hunter AL, Ferry AV, et al. High sensitivity cardiac troponin and the under-diagnosis of myocardial infarction in women: prospective cohort study. BMJ. 2015;350: g7873.
32. Bugiardini R, Ricci B, Cenko E, Vasiljevic Z, Kedev S, Davidovic G, et al. Delayed care and mortality among women and men with myocardial infarction. J Am Heart Assoc. 2017;6: e005968.
33. Kalff MC, Dijksterhuis WPM, Wagner AD, Oertelt-Prigione S, Verhoeven RHA, Lemmens V, et al. Sex differences in treatment allocation and survival of potentially curable gastroesophageal cancer: a population-based study. Eur J Cancer. 2023;187:114–23.
34. Melk A, Babitsch B, Borchert-Morlins B, Claas F, Dipchand AI, Eifert S, et al. Equally interchangeable? How sex and gender affect transplantation. Transplantation. 2019;103:1094–110.
35. Barlek MH, Rouan JR, Wyatt TG, Helenowski I, Kibbe MR. The persistence of sex bias in high-impact clinical research. J Surg Res. 2022;278:364–74.
36. Steinberg JR, Turner BE, Weeks BT, Magnani CJ, Wong BO, Rodriguez F, et al. Analysis of female enrollment and participant sex by burden of disease in US clinical trials between 2000 and 2020. JAMA Netw Open. 2021;4: e2113749.
37. Dekker M, de Vries ST, Versantvoort CHM, Drost-van Velze EGE, Bhatt M, van Meer PJK, et al. Sex proportionality in pre-clinical and clinical trials: an evaluation of 22 marketing authorization application dossiers submitted to the European Medicines Agency. Front Med (Lausanne). 2021;8: 643028.

38. Gruca TS, Hottel WJ, Comstock J, Olson A, Rosenthal GE. Sex and cardiovascular disease status differences in attitudes and willingness to participate in clinical research studies/clinical trials. Trials. 2018;19:300.
39. Matthews SJ, McCoy C. Thalidomide: a review of approved and investigational uses. Clin Ther. 2003;25:342–95.
40. Obias-Manno D, Scott PE, Kaczmarczyk J, Miller M, Pinnow E, Lee-Bishop L, et al. The food and drug administration office of women's health: impact of science on regulatory policy. J Womens Health (Larchmt). 2007;16:807–17.
41. Brady E, Nielsen MW, Andersen JP, Oertelt-Prigione S. Lack of consideration of sex and gender in COVID-19 clinical studies. Nat Commun. 2021;12:4015.
42. FDA. Precision medicine. 2018. https://www.fda.gov/medical-devices/in-vitro-diagnostics/precision-medicine. Accessed 4 Jun 2025.
43. Hood L, Friend SH. Predictive, personalized, preventive, participatory (P4) cancer medicine. Nat Rev Clin Oncol. 2011;8:184–7.
44. Heise L, Greene ME, Opper N, Stavropoulou M, Harper C, Nascimento M, et al. Gender inequality and restrictive gender norms: framing the challenges to health. Lancet. 2019;393:2440–54.
45. Greenhalgh T, Papoutsi C. Studying complexity in health services research: desperately seeking an overdue paradigm shift. BMC Med. 2018;16:95.
46. Holman D, Walker A. Understanding unequal ageing: towards a synthesis of intersectionality and life course analyses. Eur J Ageing. 2020;18:239–55.
47. Hankivsky O, Reid C, Cormier R, Varcoe C, Clark N, Benoit C, et al. Exploring the promises of intersectionality for advancing women's health research. Int J Equity Health. 2010;9:5.
48. Sapir-Pichhadze R, Oertelt-Prigione S. P3(2): a sex- and gender-sensitive model for evidence-based precision medicine: from knowledge generation to implementation in the field of kidney transplantation. Kidney Int. 2023;103:674–85.

Part II
Disease-Specific Sex Differences

Sex- and Gender-Specific Medicine of Gastric Cancer

Nayoung Kim

Abstract Gastric cancer (GC) exhibits a clear global male predominance. However, in younger patients, females are disproportionately affected and are more likely to present with diffuse or poorly differentiated tumors at advanced stages. In contrast, GC in older populations is predominantly seen in males, who often develop the intestinal subtype and exhibit a higher rate of synchronous tumors. Among females, diffuse-type GC is more common before menopause, while the frequency of intestinal-type GC gradually aligns with that of males approximately 20 years after menopause—suggesting a potential protective role of female sex hormones against intestinal-type GC development. Obesity, a known risk factor for cardia GC, shows a stronger association in males. Gastric tissue expresses sex hormone receptors such as estrogen receptors (ERα and ERβ) and the androgen receptor, with expression patterns varying according to sex, histological subtype, and *Helicobacter pylori* infection status. Epstein–Barr virus-associated GC (EBVaGC), accounting for 5–16% of all GC cases, is more frequent in males and paradoxically associated with better outcomes despite being located proximally and demonstrating poor differentiation—likely reflecting sex-based differences in immune responses. Furthermore, the therapeutic benefit of immune checkpoint inhibitors (ICIs) targeting PD-1/PD-L1 pathways has been observed predominantly in male GC patients. Altogether, these data underscore the influence of sex-specific biological factors in GC pathogenesis and treatment response. ERβ may offer protection against intestinal-type GC, and the impact of EBVaGC and ICIs appears to be amplified in males due to distinct patterns of immune regulation.

Keywords Gastric cancer · Sex · Gender · Estrogen hormone · Epstein–Barr virus

---

N. Kim (✉)
Department of Internal Medicine, Seoul National University Bundang Hospital, Seongnam, Republic of Korea
e-mail: nakim49@snu.ac.kr

Research Center for Sex- and Gender-Specific Medicine, Seoul National University Bundang Hospital, Seongnam, Republic of Korea

Department of Internal Medicine, Seoul National University College of Medicine, Seoul, Republic of Korea

H. Lee et al. (eds.), *Sex, Gender, and Emerging Technology in Healthcare: Mitigating Bias and Fostering Equity*, https://doi.org/10.1007/978-981-95-2070-1_2

1 Introduction

Recognizing the biological differences between men and women is fundamental in understanding the human body, and medical science has advanced by taking such distinctions into account. Nonetheless, many diseases still lack sufficient consideration of sex-based variations in their causes and treatments. In South Korea, gastric cancer (GC) is significantly more prevalent in men, with a male-to-female incidence ratio of approximately 2:1 [1]. Several risk factors have been implicated in the development of GC, including *Helicobacter pylori* (*H. pylori*) infection, advancing age, high dietary salt intake, limited consumption of fruits and vegetables, and exposure to alcohol and tobacco [2]. Additionally, notable sex-based differences exist in the histological types of GC. According to the Lauren classification, GC is divided into intestinal and diffuse types [3] (Fig. 1). The diffuse type is more frequently observed in females, whereas the intestinal type predominates in males. However, this classification does not fully account for the sex disparity in GC incidence. More recently, the potential involvement of sex hormones—particularly estrogen—in GC development and progression has gained attention [4]. Research has explored associations between circulating estrogen levels [4, 5] and the expression of estrogen receptors (ERs) [6–11] with GC risk and prognosis. Beyond the Lauren system, the Cancer Genome Atlas (TCGA) project (2014) proposed a molecular classification of gastric adenocarcinoma into four distinct subtypes: (1) Epstein–Barr virus (EBV)-associated gastric carcinoma (EBVaGC), (2) microsatellite instability-high tumors, (3) chromosomally unstable tumors, and (4) genomically stable tumors [12, 13]. EBV, a double-stranded DNA virus, was designated a Group I carcinogen by the International Agency for Research on Cancer (IARC) in 1997 due to its oncogenic potential [14, 15]. In South Korea, EBV is implicated in approximately 5.6–13% of GC cases [16–19]. Notably, EBVaGC occurs nearly ten times more frequently in men and has shown a more favorable prognosis exclusively in male patients [19]. In addition, EBV positivity and tumor location in the antrum have been linked to increased PD-L1 (programmed cell death ligand-1) expression in men but not in women [20]. This sex-specific pattern of PD-L1 expression parallels the findings observed in EBVaGC. This chapter focuses on the influence of sex and gender on the epidemiology, pathology, and immunological landscape of GC.

2 Incidence and Morality of Gastric Cancer

GC exhibits high incidence rates in low- and middle-income nations, with notable regional concentrations in East Asia, Eastern Europe, and parts of Central and South America [21, 22]. Asia accounts for roughly three-quarters of all global GC cases, and South Korea, in particular, reports the world's highest incidence rates for both males and females [23, 24]. In contrast, GC is relatively uncommon in North America and Northern Europe, with incidence levels comparable to those observed across African regions.

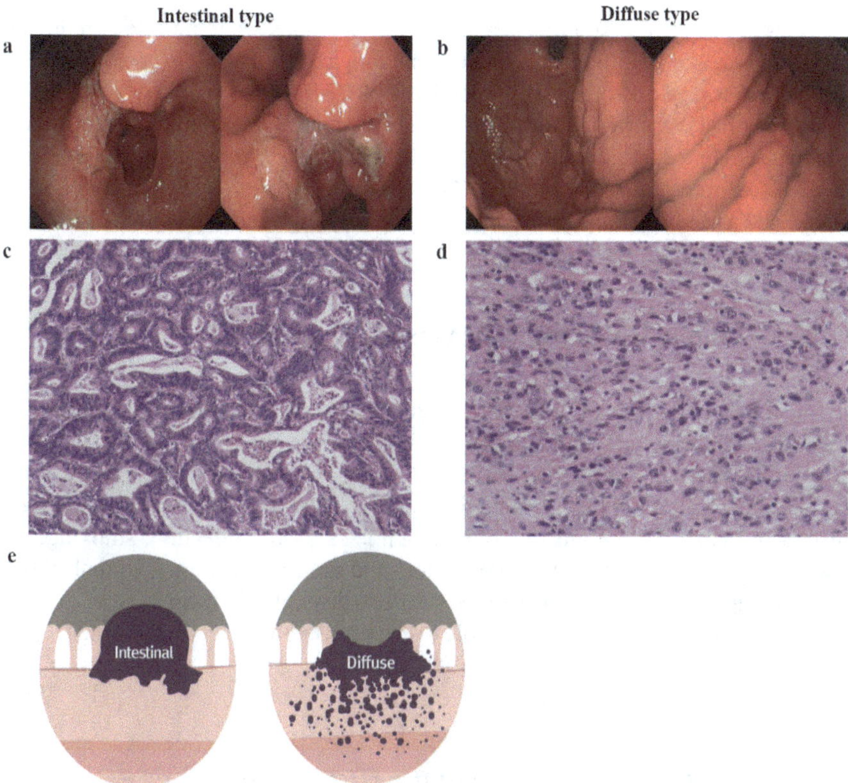

Fig. 1 Endoscopic images (**a**, **b**), histopathological features (**c**, **d**), and schematic models illustrating gastric cancer development according to histological subtype—intestinal type and diffuse type (**e**)

Evidence from migrant studies underscores the significant role of environmental influences in regional disparities in GC prevalence. For instance, first-generation Japanese immigrants to Hawaii showed lower GC incidence compared to their counterparts in Japan, and this trend was even more pronounced among second-generation, Hawaii-born Japanese individuals—though their rates remained higher than those of the native white population [25].

H. pylori infection is the predominant etiological factor for GC, particularly non-cardia GC, accounting for nearly 90% of newly diagnosed cases [26, 27]. While variations in *H. pylori* prevalence generally align with GC distribution globally, other environmental and lifestyle factors are also key contributors. These include diets high in salted foods and low in fresh fruits, as well as alcohol use and cigarette smoking—all of which are well-established risk factors [28, 29].

Although GC is often discussed as a single disease, it is commonly divided into two anatomical categories: non-cardia GC, arising in the distal stomach, and cardia GC, located near the gastroesophageal junction. Over the past 50 years, non-cardia

GC rates have declined significantly in most populations—a trend often credited to inadvertent public health improvements, including reduced *H. pylori* prevalence and better food storage and preservation methods [30, 31].

By contrast, gastric cardia cancers share more epidemiologic traits with esophageal adenocarcinoma (EAC). Key risk factors for cardia GC include obesity and gastroesophageal reflux disease (GERD), with Barrett's esophagus (BE) also considered a predisposing condition—especially in high-income countries [32].

3 Sex/Gender Differences in Gastric Cancer

To fully understand why GC occurs more frequently in males, a multifaceted analysis is essential. This includes examining genetic differences linked to sex chromosomes (X and Y), hormonal influences, lifestyle behaviors such as alcohol and tobacco use, *H. pylori* infection [33], and environmental factors within the tumor microenvironment, including the gut microbiota. While *H. pylori* infection, smoking, and alcohol use are well-known GC risk factors and are more common in males, they do not completely account for the observed sex disparity. In recent years, growing interest has focused on the potential role of estrogen exposure in modulating GC risk [34, 35].

3.1 Sex and Age Differences in the Histopathological Characteristics of Gastric Cancer

Although the global incidence of GC is generally higher in males, this pattern does not hold across all age groups [36–38]. Research examining GC's clinical and pathological characteristics by age indicates that younger patients are more frequently female and tend to present with diffuse or undifferentiated histological subtypes, often at more advanced stages of disease. In contrast, older individuals with GC are predominantly male, with a higher prevalence of the intestinal-type and a tendency to present with synchronous tumors [39–41]. As environmental conditions improve and overall GC cases decline, the relative proportion of diffuse-type GC may remain stable [42].

A study conducted at Seoul Samsung Hospital between November 2014 and May 2021 assessed 758 GC patients (531 males and 227 females) to evaluate the relationships among age, menopausal status (tracked in 5-year intervals), and tumor histology. The findings suggested that female sex hormones may play a protective role against the development of intestinal-type GC [43]. Notably, the histological pattern in women gradually approached that observed in men with increasing time since menopause. Specifically, the frequency of intestinal-type GC in premenopausal women was 19.0%, rising to 30.4% in those less than 10 years post-menopause and

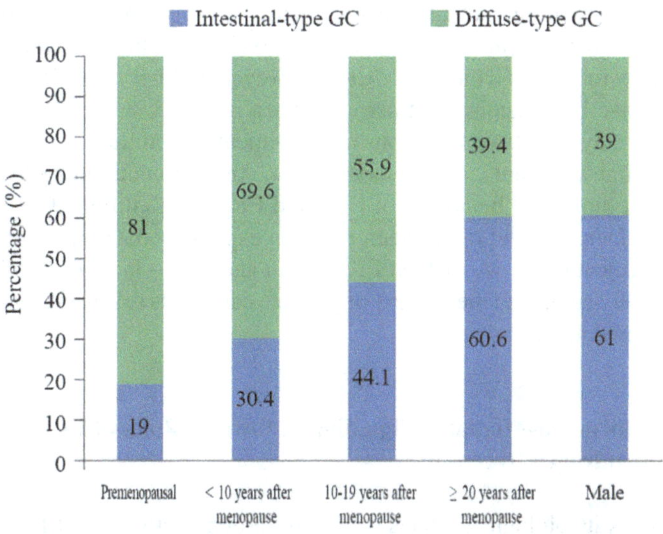

Fig. 2 The distribution of intestinal-type gastric cancer (GC) varies according to sex and menopausal status. Intestinal-type GC was significantly less frequent in premenopausal females (19.0%) and in postmenopausal females within < 10 years (30.4%) and 10–19 years (44.1%) after menopause compared to males (61%) (all $p < 0.05$). However, no significant difference was observed between males and females who were ≥ 20 years postmenopausal (60.6% vs. 61%, $p = 0.518$) (adapted from Jung et al. [44])

44.1% in those 10–19 years post-menopause. These rates were significantly lower than the 61% observed in males (all $p < 0.05$) [44] (Fig. 2). However, among females more than 20 years post-menopause, the proportion of intestinal-type GC was nearly identical to that in men (60.6 vs. 61%, $p = 0.518$) [44].

These shifts in the distribution of GC histological types over time suggest that estrogen may exert a protective effect against intestinal-type GC [43–45]. Thus, younger females are significantly less likely to develop intestinal-type GC compared to males, but this difference diminishes with age and is no longer apparent approximately 20 years after menopause—around the age of 70 [46].

3.2 Hormonal Influence on Gastric Carcinogenesis: Estrogen and Androgen Receptors

Gastric tissues have been found to independently express four types of sex hormone receptors: ER alpha (ERα), ER beta (ERβ), progesterone receptor (PR), and androgen receptor (AR), indicating a potential involvement of these receptors in gastric tumorigenesis [11]. Numerous studies have attempted to elucidate the observed sex differences in GC by focusing on the role of ERs. However, findings regarding ER expression in GC have been inconsistent across studies [6, 9, 33, 36, 45, 47–49]. These discrepancies may stem from variations in experimental conditions, including differences in cell lines, reagents, and methodologies. Another contributing factor could be the distinct biological behaviors of ERα and ERβ across different GC histological types. In fact, the field of estrogen research experienced a major turning point with the identification of two distinct ERs—ERα and ERβ—in 1996. While AR has been extensively studied in the context of prostate cancer, its role in GC has received relatively limited attention.

3.2.1 Sex Hormone–Mediated Signaling Pathways: Roles of Estrogen and Androgen Receptors

Estrogen exerts its biological effects primarily through activation of ERs [9]. ERα and ERβ belong to the nuclear receptor superfamily, which enables the conversion of extracellular cues into gene transcription responses [50–52] (Fig. 3a). Genomic signaling occurs when estrogen binds to ERs and the complex interacts with estrogen response elements (EREs) within the promoters of target genes. Estrogen can also induce non-genomic effects by binding to G-protein coupled ER 1 (GPER1), which subsequently influences ion channel activity or activates downstream signaling cascades such as PI3K/Akt and MAPK pathways [53] (Fig. 3b).

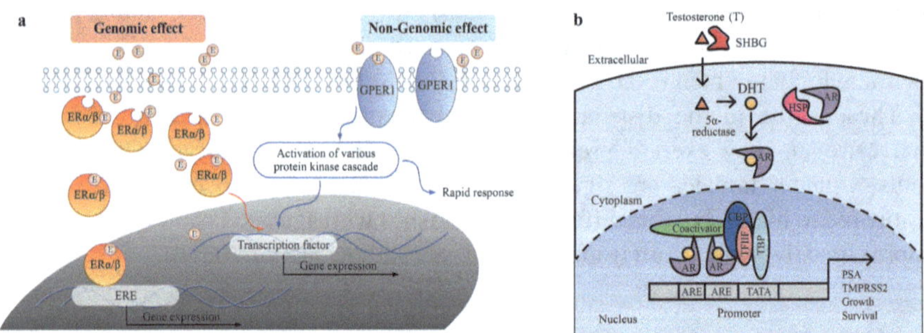

Fig. 3 Estrogen and androgen receptor pathways. (**a**) Estrogen exerts genomic effects by binding to nuclear estrogen receptors (ERs), which interact with estrogen response elements (EREs) in the promoter regions of target genes. Non-genomic effects are mediated through the membrane-bound G protein-coupled estrogen receptor 1 (GPER1), leading to modulation of ion channels and activation of intracellular signaling cascades, including the PI3K/Akt and MAPK pathways (adapted from Chen et al. [52]). (**b**) Testosterone is enzymatically converted to dihydrotestosterone (DHT), which binds to the androgen receptor (AR). The AR–DHT complex translocates into the nucleus and binds to androgen response elements (AREs) in the promoter regions of target genes, initiating transcriptional activity. This signaling cascade is analogous to the estrogen pathway (adapted from Tan et al. [53])

In the case of androgens, testosterone is metabolized into dihydrotestosterone (DHT), which binds to the AR. The DHT-AR complex then translocates into the nucleus and interacts with androgen response elements (AREs) in the promoter regions of specific genes—an action mechanistically similar to that of estrogen signaling.

Matsuyama et al. examined the immunohistochemical staining (IHC) patterns of ERs in both cancerous and non-cancerous gastric tissues [54]. They reported that ERβ was expressed in approximately 90% of cases, and that staining intensity and distribution varied according to tumor histology and patient age. Another study found ERβ expression in about 45% of intestinal-type GC samples, whereas diffuse-type GC cases showed no ERβ positivity, and ERα was consistently negative [6]. In a separate investigation from China, ERα, ERβ, and AR were detected in 6% (9/150), 93.5% (143/153), and 42.4% (59/139) of GC tissues, respectively [7]. However, this study did not account for histological subtypes based on Lauren classification, nor did it detail staining patterns of these receptors [7]. Given the variability in findings across studies, further research is warranted to clearly define the expression profiles and regulatory roles of ERβ in both malignant and non-malignant gastric tissues, especially in relation to sex-specific differences [55].

3.2.2 Sex Differences in Sex Hormone Receptor Expression and Their Clinicopathological Implications in Gastric Cancer

Although numerous investigations have sought to clarify the male predominance in GC, many experimental and clinical studies still contain inherent biases. For example, in vitro experiments often overlook the sex of cell lines, and only around 25% of GC cell lines have documented sex information. Furthermore, many clinical studies have failed to treat sex as a key variable, resulting in a lack of data on sex hormone exposure levels. While some reports have explored GC prognosis by evaluating differences in pathology and histology across age and sex groups [56, 57], these investigations often introduce selection bias by restricting the treatment types analyzed, limiting their generalizability to the broader GC patient population. Accordingly, large-scale or prospective studies may be more suitable for validating the hypothesis that sex and gender influence GC risk. For instance, analyses using data from the Surveillance, Epidemiology, and End Results (SEER) program—an extensive cancer registry supported by the U.S. National Cancer Institute—have demonstrated improved overall and cancer-specific survival in female GC patients [58].

ERs are detectable in 2–30% of human GC cases, especially in poorly differentiated tumors [59]. Several studies have linked sex hormone receptor expression to GC prognosis [6–11]. Specifically, ER-positive GC cases exhibit a significantly higher frequency of undifferentiated histology compared to differentiated tumors [33], and further studies have also connected ER status to tumor stage [34, 35]. For example, Yi et al. reported that ERα expression was correlated with diffuse-type GC and shorter disease-free survival [36]. In diffuse-type GC patients with overt *H. pylori* infection, ERα expression was also associated with greater tumor invasiveness and higher HOTAIR expression levels [60].

Conversely, well-differentiated gastric adenocarcinomas typically show higher ERβ expression, while poorly differentiated tumors are often marked by reduced or absent ERβ levels [55, 61]. ERs exist in multiple isoforms [9], with ERβ generally exhibiting tumor-suppressive functions [8–10], whereas ERα is more commonly linked to tumor progression [7, 9, 10]. Mechanistically, ERα may promote cancer by regulating tumor suppressors and cell cycle proteins such as p53, p21, p27, cyclin D1, and E-cadherin [7], as well as modulating signaling via the c-Src pathway [62] and glucose-regulated protein 94 [63]. In contrast, ERβ appears to counteract tumor progression by inhibiting cell proliferation and invasion through negative regulation of the mTOR–Arpc1b/EVL pathway [8]. Likewise, AR is believed to be associated with aggressive GC behavior, akin to ERα, and has been linked to enhanced cell motility and invasiveness [7].

Beyond GC, ERβ has also demonstrated protective roles in other malignancies, such as colorectal cancer (CRC) [64, 65]. In our dataset, ERα and AR were both upregulated in GC tissues (Fig. 4a and e) relative to non-cancerous tissues (Fig. 4b and f), and their expression was associated with diffuse-type and poorly differentiated histology, suggesting a relationship with poorer clinical outcomes [55]. Conversely, ERβ expression showed an opposite pattern (Fig. 4c and d), consistent with most prior findings [55]. RT-PCR analyses further indicated lower ERβ levels in GC tissues than in controls, supporting a possible protective role of ERβ against tumor development [55].

Subgroup analysis based on IHC provided further insights [55]. ERα expression scores were significantly elevated in cancer tissues ($p < 0.001$), particularly in cardia GC ($p < 0.001$) and undifferentiated tumors ($p = 0.015$). ERα expression was also marginally associated with *H. pylori*-negative status ($p = 0.081$) and diffuse or mixed histology ($p = 0.061$) [55]. ERβ expression, by contrast, was more frequent in *H. pylori*-positive cases ($p = 0.005$) and marginally associated with alcohol use ($p = 0.087$) and undifferentiated tumors ($p = 0.079$) [55]. AR scores were significantly higher in cancers ($p < 0.001$), especially cardia GC ($p < 0.001$), diffuse or mixed types ($p = 0.012$), and undifferentiated cancers ($p = 0.001$). In both cancer and control groups, ERα and AR showed similar expression trends, while ERβ displayed an inverse pattern [55] (Fig. 5).

These findings suggest that ERα, ERβ, and AR expression levels differ by sex, histological subtype, and *H. pylori* infection status—factors that may help explain sex-specific GC patterns [55]. Emerging research also indicates that AR contributes to cancer development in tissues beyond the reproductive system. AR overexpression is associated with lymph node metastasis and worse outcomes [38, 39]. In patients with AR-positive GC tissues, AR was not linked to other clinical parameters but independently predicted lower survival [37, 40, 41]. Notably, intestinal-type GC stroma in male patients had a significantly higher AR positivity rate compared to females, possibly contributing to increased tissue permeability and invasiveness in men. This opens the door for potential use of AR-targeted therapies in GC [66].

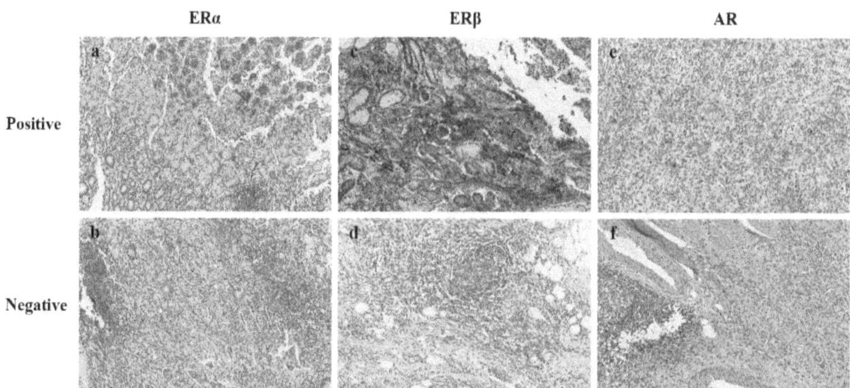

Fig. 4 Representative immunohistochemical staining of sex hormone receptors. (**a**, **b**) Estrogen receptor alpha (ERα); (**c**, **d**) estrogen receptor beta (ERβ); (**e**, **f**) androgen receptor (AR). In all panels except (**d**), representative images of both positively and negatively stained slides are shown. In panel (**d**), a negative control image is presented in which only a secondary antibody was applied—without a primary antibody—on a slide that was ERβ-positive in panel (**c**), to illustrate the non-specific staining pattern observed with ERβ. ERs exhibited non-specific staining not only in the nuclei of tumor cells but also in surrounding connective tissue and blood vessels (**a**, **c**), whereas AR staining was predominantly nuclear and confined to tumor cells (**e**) (adapted from Choi et al. [55])

Additionally, male patients—particularly those with stage III GC—exhibited higher recurrence and liver metastasis rates than females. This difference may be partly explained by increased PD-L1 expression in males and older patients [67]. Altogether, these observations support the hypothesis that sex hormones play a fundamental role in the sex-related disparities observed in GC [68, 69].

4 Gastric Cancer and Body Weight in Term of Sex

While excessive body weight is a well-established risk factor for various malignancies—including CRC, esophageal adenocarcinoma, kidney cancer, pancreatic cancer, and breast cancer [70–72]—the association between obesity and GC remains less well defined. Furthermore, the relationship between body weight and GC risk appears to vary depending on the anatomical subsite of the tumor. GC is broadly categorized into two subtypes based on tumor location: cardia and non-cardia GC, each characterized by distinct etiological and pathological features [73]. Notably, emerging evidence suggests that underweight status may also be linked to an elevated risk of GC, particularly in East Asian populations, where non-cardia GC comprises more than 90% of total GC cases [74–77].

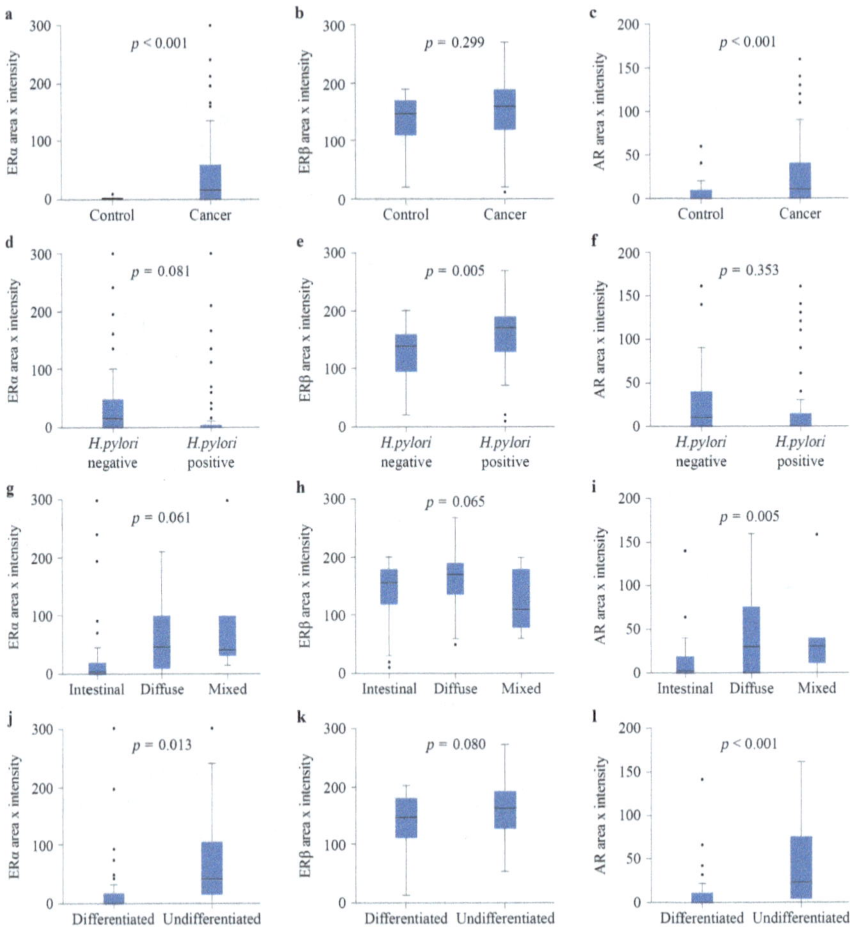

Fig. 5 Expression scores of sex hormone receptors based on immunohistochemistry. (**a–c**) Comparison of expression scores between control and gastric cancer (GC) groups; (**d–f**) expression scores according to *Helicobacter pylori* (*H. pylori*) infection status; (**g–i**) expression scores by Lauren classification (intestinal vs. diffuse); (**j–l**) expression scores by World Health Organization (WHO) histological classification (adapted from Choi et al. [55])

4.1 Overweight/Obesity in Gastric Cancer

The association between obesity and GC incidence appears to differ according to the tumor's anatomical location within the stomach [78]. Obesity has been consistently linked to a heightened risk of cardia GC, with significantly increased incidence observed among obese individuals compared to those with normal body weight. In contrast, current evidence suggests no definitive association between obesity and non-cardia GC [78]. Recent studies have further indicated that the impact of obesity on GC risk may be modulated by sex [78]. Notably, an increased risk of GC has also been observed among underweight females, particularly in East Asian populations.

4.1.1 Association Between Overweight/Obesity and Gastric Cancer Incidence

The global surge in obesity over recent decades has affected both men and women, contributing to the rising burden of numerous chronic diseases, including type 2 diabetes, cardiovascular disease, stroke, and various malignancies [79–81]. Among these, gastric cardia cancer—located at the proximal end of the stomach—has demonstrated a particularly strong association with excess body weight. Both the World Cancer Research Fund (WCRF) and the International Agency for Research on Cancer (IARC) have identified overweight and obesity as well-established risk factors for cardia GC, supported by robust epidemiological evidence [81, 82].

A recent systematic review confirmed this link, revealing that individuals with a body mass index (BMI) $\geq$ 25 kg/m^2 (overweight) and $\geq$ 30 kg/m^2 (obese) had significantly elevated risks of developing cardia GC, with hazard ratios (HRs) of 1.21 (95% CI, 1.03–1.42) and 1.82 (95% CI, 1.32–2.49), respectively. In contrast, no significant association was found between obesity and non-cardia GC (HR 1.00; 95% CI, 0.87–1.15) [78]. The estimated population attributable fraction for cardia GC due to obesity was 8.8% in males (95% CI, 3.0–14.8%) and 11.2% in females (95% CI, 3.8–18.8%), indicating that nearly one in ten cases of cardia GC may be attributable to obesity in both sexes [83].

To further explore sex-specific differences, a large-scale cohort study using the National Health Insurance Service–Health Screening Cohort in Korea analyzed data from 333,169 adults followed over a median of 12 years [84]. Participants were stratified into five BMI categories: <18.5, 18.5–22.9, 23.0–24.9, 25.0–29.9, and $\geq$30.0 kg/m^2. Cox proportional hazards models showed that, in men, obesity (BMI $\geq$ 30.0 kg/m^2) was associated with a 1.27-fold increased risk of GC (95% CI, 1.02–1.57), while no significant association was observed in women [84]. For non-cardia GC, a U-shaped risk pattern emerged in males, with a significantly reduced risk noted in the 23.0–24.9 kg/m^2 BMI range. In women, however, a larger waist circumference—a marker of visceral adiposity—was significantly associated with increased GC risk (HR 1.37; 95% CI, 1.07–1.74) [84]. These findings suggest that maintaining a BMI in the 23.0–24.9 kg/m^2 range may minimize non-cardia GC risk in men, while visceral obesity represents a more critical risk factor for women.

Mechanistically, one plausible explanation for the obesity–cardia GC link is the increased prevalence of gastroesophageal reflux disease (GERD) among obese individuals [85–87]. Abdominal obesity elevates intra-abdominal pressure, promoting gastric reflux and increasing the risk of GERD [88]. Numerous observational studies have confirmed this association, and a meta-analysis demonstrated a significant relationship between overweight/obesity and GERD (OR 1.66; 95% CI, 1.61–1.71), albeit with some heterogeneity across studies [89].

GERD comprises a spectrum of clinical manifestations, with erosive esophagitis—marked by mucosal damage—being particularly concerning due to its potential progression to Barrett's esophagus (BE), a precursor of adenocarcinoma at the esophagogastric junction. It is estimated that 1% to 13% of patients with erosive esophagitis develop BE [90–92], suggesting a potential carcinogenic sequence from obesity to GERD to BE, ultimately leading to cardia cancer.

In addition to mechanical factors like GERD, obesity-associated biological pathways may also contribute to carcinogenesis. Metabolic alterations, including hyperinsulinemia and elevated insulin-like growth factors (IGFs), may enhance cellular proliferation and suppress apoptosis. Moreover, pro-inflammatory adipokines—such as leptin, adiponectin, tumor necrosis factor-alpha (TNF-α), and interleukin-6 (IL-6)—produced by adipose tissue in obese and diabetic states can promote chronic inflammation and tumor development [93, 94].

4.1.2 Impact of Sex on the Obesity–Cardia Gastric Cancer Relationship

Researchers have explored whether the elevated risk of GC associated with overweight and obesity varies by sex. A systematic review reported that overweight or obese men had a significantly higher risk of GC compared to normal-weight men (hazard ratio [HR], 1.12; 95% confidence interval [CI], 1.00–1.24), whereas no significant risk difference was observed in women (HR, 1.04; 95% CI, 0.93–1.16) [78]. In South Korea, one study demonstrated a significant association between obesity and early-stage GC in males, while in females, obesity was more closely linked to gastric dysplasia, a recognized precursor to GC [95]. However, most studies did not differentiate GC cases by anatomical site, limiting conclusions regarding sex-specific differences in the obesity–cardia GC relationship. Nonetheless, given the established association between obesity and cardia GC, along with the higher GC incidence in men, it is plausible that obesity contributes more substantially to the risk of cardia GC in males [78].

The nearly twofold higher incidence of GC in males has been partially attributed to the protective effects of female sex hormones. Multiple studies have found reduced GC risk in females with longer reproductive periods, use of hormone replacement therapy, or higher parity [43, 96]. Furthermore, obesity in postmenopausal females has been associated with elevated circulating levels of estrone, estradiol, and free estradiol [97–102]. These hormonal alterations may counterbalance the detrimental effects of obesity on GC risk, potentially explaining the weaker association observed in women. Supporting this hypothesis, a recent study found that obesity was significantly associated with GC only in postmenopausal females, with no correlation identified in premenopausal females—highlighting the protective role of estrogen [103].

Erosive esophagitis, a known risk factor for Barrett's esophagus (BE), is more prevalent in men, with a pooled male-to-female prevalence ratio of 1.57 (95% CI, 1.40–1.76) [104]. In South Korea, this ratio is even higher at approximately 3.5:1 [105]. BE and esophagogastric adenocarcinoma (EAC), both of which demonstrate strong male predominance, may further explain the heightened risk of cardia GC in obese males [106, 107].

Interestingly, the sex gap in gastroesophageal reflux disease (GERD) narrows with age. Postmenopausal women exhibit increased rates of erosive esophagitis, likely due to estrogen decline [108]. Estrogen has been shown to enhance esophageal barrier integrity by upregulating tight junction proteins, offering a potential mechanistic explanation for its protective effect [109].

Additionally, obesity leads to elevated serum leptin levels, and higher leptin concentrations have been linked to increased BE risk—particularly in men [110]. These findings suggest that obesity-related mechanisms, including GERD and hormonal factors such as leptin, may exert a more significant impact on cardia GC development in men than in women.

4.2 *Underweight in Gastric Cancer*

Recent studies have demonstrated that individuals with a low body mass index (BMI) exhibit a significantly increased risk of developing GC compared to those with normal BMI, particularly in East Asian populations where non-cardia GC comprises a substantial proportion of total GC cases [74–77, 111–113]. These findings support a strong association between underweight status and an elevated susceptibility to non-cardia GC.

4.2.1 Association Between Underweight and Non-cardia Gastric Cancer Incidence

One possible explanation for the lack of association between underweight status and increased non-cardia GC risk in earlier studies is the geographic and demographic context in which these studies were conducted. Most prior investigations were based in Western populations, where non-cardia GC comprises a relatively small proportion of overall GC cases compared to Asian populations [113]. Additionally, Western cohorts typically have lower rates of underweight individuals and higher rates of obesity than their Asian counterparts [79], potentially limiting the statistical power to detect a meaningful association between low body mass index (BMI) and non-cardia GC. Moreover, earlier studies often focused predominantly on the effects of overweight and obesity on GC risk, frequently excluding underweight individuals or grouping them with those of normal BMI. This methodological limitation may have obscured any independent effects of underweight status on GC risk.

Although a definitive causal relationship between low BMI and non-cardia GC has yet to be fully established, several plausible biological mechanisms may underlie this association. Underweight status is often indicative of chronic malnutrition, which can result in deficiencies in micronutrients with known antioxidant and cancer-protective properties [114, 115]. Furthermore, individuals with low body weight may experience compromised immune function due to nutritional deficiencies [116, 117], potentially increasing their susceptibility to chronic infections such as *H. pylori*, a well-established carcinogenic agent for non-cardia GC.

In our recent investigation, we identified a sex-specific association between BMI and survival outcomes in different GC subtypes [118]. Among males with cardia GC, a U-shaped relationship was observed between BMI and mortality: underweight (9.6%), normal weight (6.4%), overweight (6.1%), obese (5.6%), and severely obese (9.3%) (Fig. 6a). This pattern was not observed in females ($p = 0.003$) (Fig. 6b). In female patients, the prevalence of diffuse-type GC decreased with increasing BMI: underweight (59.9%), normal (56.8%), overweight (49.5%), obese (44.8%), and severely obese (41.7%) (Fig. 6d), whereas no such trend was evident in males ($p < 0.001$) (Fig. 6c). In both sexes, underweight status was associated with the poorest prognosis ($p < 0.001$). However, higher BMI was correlated with improved survival only in males [118]. These sex-based differences in prognosis by BMI were particularly pronounced in subgroup analyses stratified by TNM stages I to III and in patients who underwent surgical resection [118].

4.2.2 Sex-Based Differences in the Association Between Underweight and Non-cardia Gastric Cancer

To date, only a limited number of studies have indicated an increased risk of GC among individuals with low body weight, and most of these have been published in recent years. Consequently, whether sex-based differences influence the relationship between underweight status and GC risk remains uncertain.

Among the existing literature, only three observational studies have investigated potential sex-specific differences in the association between underweight and GC risk—or conversely, between higher BMI and reduced GC risk [74, 75, 111]. One study conducted in a Western population found that underweight females had a greater increase in GC risk compared to males; however, this difference was not statistically significant [111]. Another study involving East Asian participants reported a similar magnitude of increased GC risk in both sexes when individuals with a BMI of 20.32–21.76 kg/m^2 were compared to those with a BMI of 23.31 kg/m^2 [74]. A third study observed a possible association between underweight and an elevated risk of non-cardia GC in females, though statistical significance was not reached [75]. Thus, current evidence is insufficient to determine whether sex meaningfully modifies the association between underweight and GC risk.

Nonetheless, biological mechanisms may provide insight into how this risk might differ by sex. In postmenopausal women with a BMI below 22.5 kg/m^2, serum levels of sex hormones such as estradiol and estrone are significantly reduced [119, 120]. Conversely, prior studies have shown that high-dose estrogen therapy is associated with a decreased standardized incidence ratio of GC [121], and that diffuse-type GC—which is more prevalent in women—is closely associated with ER positivity

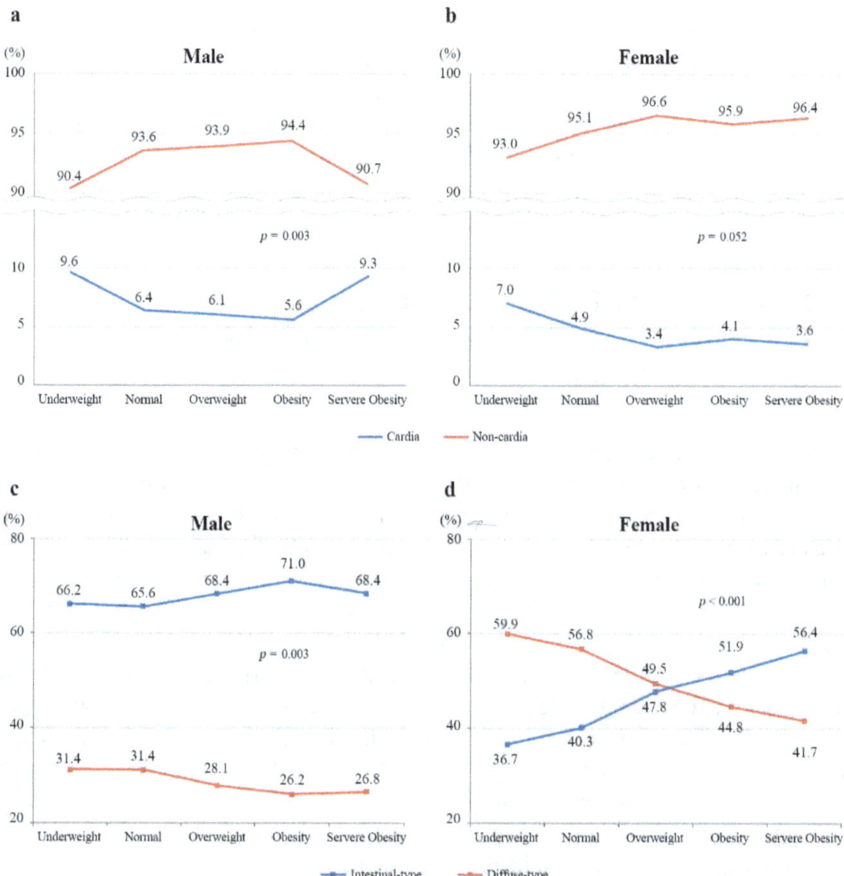

Fig. 6 Sex differences in the distribution of gastric cancer (GC) by tumor location and histological type according to body mass index (BMI). (**a**) In males, a U-shaped relationship was noted, with a higher proportion of cardia GC in both the underweight and severely obese groups. (**b**) In contrast, among females, only the underweight group demonstrated a higher proportion of cardia GC. (**c**) The distribution of non-cardia GC in males remained relatively stable across all BMI categories. (**d**) In females, a distinct shift in histological subtype was observed across BMI strata: in the underweight group, intestinal-type and diffuse-type GC accounted for 36.7% and 59.9%, respectively, whereas in the severely obese group, these proportions were reversed—56.4% for intestinal-type and 41.7% for diffuse-type GC. BMI categories were defined according to the Asia–Pacific World Health Organization criteria: underweight (<18.5 kg/m^2), normal weight (18.5–22.9 kg/m^2), overweight (23.0–24.9 kg/m^2), obesity (25.0–29.9 kg/m^2), and severe obesity (≥30.0 kg/m^2) (adapted from Jo et al. [118])

[36]. Furthermore, both animal and cellular models have demonstrated that estrogen exerts a protective effect against gastric carcinogenesis [122, 123].

Collectively, these findings suggest that underweight females may experience a decline in protective estrogenic effects compared to those with normal or higher BMI. This hormonal deficiency could contribute to increased vulnerability to GC in underweight women.

5 Sex-Dependent Different Clinicopathological Characterization of Epstein–Barr Virus-Associated Gastric Carcinoma

EBV, a double-stranded DNA virus of the Herpesviridae family, infects epithelial cells in the oropharynx and is primarily spread via saliva. Initial infection typically occurs during childhood and is often asymptomatic. After the primary infection, the virus remains latent in B lymphocytes. Globally, over 90% of adults show seropositivity for EBV antibodies, indicating past exposure [124]. Due to its established oncogenic potential, the International Agency for Research on Cancer (IARC) classified EBV as a Group I carcinogen in 1997 [14, 15].

GC exhibits notable sex-related differences; diffuse-type GC is more prevalent in females, while GC in males tends to occur more frequently in the antrum [44, 46]. Moreover, some research has indicated that females may exhibit more active immune responses in the context of GC. As EBV-associated GC (EBVaGC) is known to be immunogenic, its biological behavior may vary by sex. Nevertheless, most existing studies on this topic have involved relatively small cohorts, and findings from meta-analyses have often been inconsistent.

5.1 The Characteristics of Epstein–Barr Virus-Associated Gastric Carcinoma

The definitive method for diagnosing EBVaGC is in situ hybridization (ISH) targeting EBV-encoded RNA in gastric tissue specimens [19, 125] (Fig. 7). Since Burke and colleagues first identified EBVaGC in 1990 [126], numerous studies have explored this distinct subtype of GC. Recent estimates suggest that EBVaGC represents the most prevalent malignancy linked to EBV, with approximately 75,000 to 90,000 new cases worldwide each year. This accounts for roughly 10% of all GCs, with a notably higher prevalence in East Asian countries, where overall GC incidence is elevated [127]. In South Korea specifically, EBV is implicated in about 5.6–13% of GC cases [16–18].

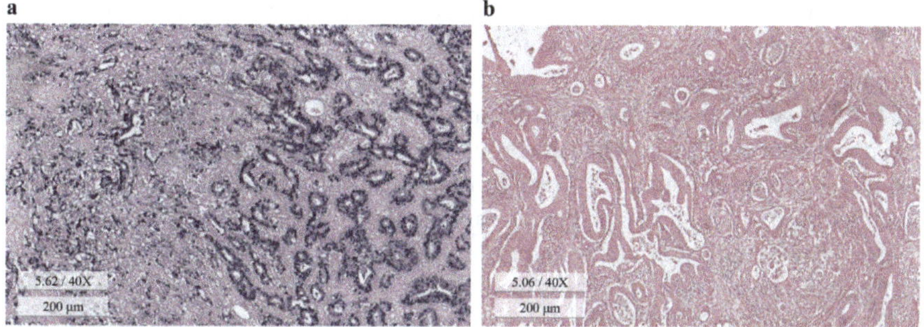

Fig. 7 Image of Epstein–Barr virus (EBV) in situ hybridization (EBV-ISH) status. (**a**) EBV-positive case showing strong nuclear staining indicative of EBV presence. (**b**) EBV-negative case in which no nuclear staining was observed on EBV-ISH (adapted from Kim et al. [19])

EBVaGC displays several characteristics that distinguish it from conventional GC. It has been reported to occur more frequently in males and younger individuals, although some meta-analyses have found no statistically significant age-related difference [128–132]. Cigarette smoking is recognized as a contributing factor, with current smokers showing a 2.4-fold higher risk and former smokers exhibiting a two-fold increased incidence compared to never-smokers [133].

The association of EBVaGC with Lauren histological classification remains inconclusive. While some studies suggest a predominance of the intestinal type, others have observed a higher frequency of the diffuse type or reported no clear correlation between EBV status and histological subtype [129, 130, 134, 135]. According to the World Health Organization (WHO) classification, EBVaGC is frequently linked to poorly differentiated carcinoma [17]. Anatomically, it most often arises in the proximal region of the stomach [17, 129].

Additionally, some investigations have indicated a weak or absent association between EBVaGC and *H. pylori* infection [136]. Compared to EBV-negative GC, EBVaGC tends to exhibit fewer lymph node metastases and reduced vascular invasion, contributing to a more favorable prognosis [128, 131, 137–140]. This improved clinical outcome may be due to its strong immunogenic nature, which elicits a robust host immune response against the tumor [141–143].

5.2 Sex Difference of Epstein–Barr Virus-Associated Gastric Carcinoma and Its Effect on the Survival of Patients with Gastric Cancer

Multiple studies have indicated that EBV-associated GC (EBVaGC) is more frequently observed in males and tends to occur at a younger age. Nevertheless, some meta-analyses have failed to confirm age as a significant factor in EBVaGC development [128–132]. Smoking has been consistently identified as a risk factor, with current smokers having a 2.4-fold increased risk and former smokers exhibiting a two-fold higher risk compared to never-smokers [133].

In our study analyzing the clinicopathological features and prognostic outcomes of EBVaGC, we explored sex-specific differences in a cohort of 4,587 GC patients who underwent EBV in situ hybridization (EBV–ISH) [19] (Table 1). In both male and female patients with EBV-positive tumors, the most frequent tumor site was the upper third of the stomach [19]. Regarding histological classification by Lauren, a significant finding was noted among males: the EBV-positive group (n = 404) had a greater proportion of diffuse-type cancers than the EBV-negative group (n = 2,625). However, this difference was not observed in females, where the proportion of diffuse-type tumors in EBV-positive (n = 52) and EBV-negative (n = 1,506) groups was comparable [19].

From an immunogenetic perspective, p53 mutations were significantly less frequent in EBV-positive males (11.7%) than in EBV-negative males (36.6%), while this difference was not significant in females (16.0 vs. 21.1%) [19] (Table 1). Microsatellite instability (MSI), however, did not vary by EBV status in either sex, suggesting that EBV-associated immune responses may not be linked to MSI in GC.

Additionally, among males, early GC (EGC) was significantly more common in the EBV-positive group (62.4%) than in the EBV-negative group (56.7%), a trend not seen in females (61.5 vs. 62.0%) [19] (Table 1). Cancer staging and the presence of lymphatic or vascular invasion also reflected this sex difference: male EBV-positive patients presented with earlier-stage disease and less invasion compared to their EBV-negative counterparts, while no such differences were found among females [19].

Survival outcomes further highlighted sex-based discrepancies. Kaplan–Meier analyses revealed that overall survival (Fig. 8a) and GC-specific survival (Fig. 8b) were both significantly better in EBV-positive males than in EBV-negative males [19]. Specifically, the 5-year cumulative overall survival was 87.1% in EBV-positive males compared to 81.4% in EBV-negative males ($p = 0.004$) (Fig. 9a), while the 5-year cancer-specific survival was 90.8% vs. 85.3%, respectively ($p = 0.003$) (Fig. 9b). In contrast, no significant survival differences were detected in females between the EBV-positive and EBV-negative groups (overall survival: $p = 0.96$; cancer-specific survival: $p = 0.90$) [19] (Fig. 9c, d).

Next, we conducted sex-stratified analyses to identify factors associated with overall survival and GC-specific survival using both univariate and multivariate Cox regression models, with the results presented in Table 2 [19]. Variables included in the univariate analysis were age, smoking history, tumor size, tumor location, histologic subtype, Lauren classification, TNM stage, and EBV–ISH status [19]. In the multivariate analysis, EBV positivity emerged as an independent favorable prognostic factor for GC-specific survival in males (HR = 0.517, $p = 0.005$), but not in females [19] (Table 2).

EBV was the first virus linked to human cancers and is implicated in several lymphoid and epithelial malignancies, including Burkitt lymphoma, Hodgkin and non-Hodgkin lymphomas, and nasopharyngeal carcinoma. Many of these are closely related to immune cell function. Sex hormones are known to influence immune

Table 1 Baseline characteristics of gastric cancer patients according to EBV positivity by sex

Variable	Male (n = 3,029)			Female (n = 1,558)		
	EBV-positive (n = 404) (13.3%)	EBV-negative (n = 2,625) (86.7%)	p- value	EBV-positive (n = 52) (3.3%)	EBV-negative (n = 1,506) (96.7%)	p- value
Age (mean ± SD)	61.0 (±10.1)	62.1 (±11.7)	0.054	63.0 (±11.5)	59.4 (±13.6)	**0.03**
≤60	200 (49.5)	1,188 (45.3)	0.111	20 (38.5)	777 (51.6)	0.063
>60	204 (50.5)	1,437 (54.7)		32 (61.5)	729 (48.4)	
Smoking	(n = 3,015)		**0.046**	(n = 1,551)		0.620
Non-smoker	90 (22.4)	710 (27.2)		47 (90.4)	1,383 (92.3)	
Smoker	311 (77.6)	1,904 (72.8)		5 (9.6)	116 (7.7)	
Size (cm)	(n = 2,864)			(n = 1,477)		
	3.7 (±2.7)	4.0 (±3.0)	0.065	3.1 (±1.9)	3.8 (±3.1)	**0.018**
≤3	211 (54.9)	1,223 (49.3)	**0.041**	25 (50.0)	754 (52.8)	0.693
>3	173 (45.1)	1,257 (50.7)		25 (50.0)	673 (47.2)	
Location	(n = 3,026)		**<0.001**	(n = 1,554)		**<0.001**
Upper	208 (51.5)	502 (19.1)		28 (53.8)	283 (18.8)	
Middle	115 (28.5)	512 (19.5)		16 (30.8)	412 (27.4)	
Low	81 (20.0)	1,608 (61.3)		8 (15.4)	807 (53.7)	
Histology			**<0.001**			**<0.001**
Differentiated	112 (27.7)	1,334 (50.8)		11 (21.2)	436 (29.0)	
Undifferentiated	202 (50.1)	902 (34.4)		24 (46.2)	783 (52.0)	
Mixed carcinoma	24 (5.9)	286 (10.9)		5 (9.6)	260 (17.3)	
Others	66 (16.3)	103 (3.9)		12 (23.0)	27 (1.7)	
Lauren classification	(n = 2,867)		**<0.001**	(n = 1,479)		**<0.001**
Intestinal	197 (51.6)	1,584 (63.7)		18 (36.0)	557 (39.0)	
Diffuse	157 (41.1)	818 (32.9)		27 (54.0)	813 (56.9)	
Mixed	7 (1.8)	64 (2.6)		3 (6.0)	56 (3.9)	
Indeterminate	21 (5.5)	19 (0.8)		2 (4.0)	3 (0.2)	
p53 mutation	(n = 1,894)		**<0.001**	(n = 992)		0.537
Negative	173 (88.3)	1,077 (63.4)		21 (84.0)	763 (78.9)	
Positive	23 (11.7)	621 (36.6)		4 (16.0)	204 (21.1)	
MSI	(n = 2,635)		**<0.001**	(n = 1,350)		**0.024**
MSS, MSI-L	291 (99.3)	2,120 (90.5)		35 (100.0)	1,147 (87.2)	
MSI-H	2 (0.7)	222 (9.5)		0 (0.0)	168 (12.8)	

(continued)

Table 1 (continued)

Variable	Male (n = 3,029)			Female (n = 1,558)		
	EBV-positive (n = 404) (13.3%)	EBV-negative (n = 2,625) (86.7%)	p- value	EBV-positive (n = 52) (3.3%)	EBV-negative (n = 1,506) (96.7%)	p- value
Cancer type			**0.032**			0.944
Early gastric cancer	252 (62.4)	1,489 (56.7)		32 (61.5)	934 (62.0)	
Advanced gastric cancer	152 (37.6)	1,136 (43.3)		20 (38.5)	572 (38.0)	
TNM staging			**0.001**			0.528
1	280 (69.3)	1,555 (59.3)		36 (69.2)	950 (63.1)	
2	44 (10.9)	416 (15.8)		8 (15.4)	217 (14.4)	
3	56 (13.9)	485 (18.5)		7 (13.5)	240 (15.9)	
4	24 (5.9)	169 (6.4)		1 (1.9)	99 (6.6)	
Lymphatic invasion	(n = 2,866)		**<0.001**	(n = 1,479)		0.399
Absent	291 (76.2)	1,531 (61.6)		37 (74.0)	977 (68.4)	
Present	91 (23.8)	953 (38.4)		13 (26.0)	452 (31.6)	
Vascular invasion	(n = 2,866)		**0.001**	(n = 1,479)		0.267
Absent	354 (92.7)	2,154 (86.7)		48 (96.0)	1,309 (91.6)	
present	28 (7.3)	330 (13.3)		2 (4.0)	120 (8.4)	
Treatment			0.084			0.685
Endoscopic submucosal dissection	11 (2.7)	118 (4.5)		1 (1.9)	41 (2.7)	
Operation	363 (89.9)	2,242 (85.4)		49 (94.2)	1,342 (89.1)	
Chemotherapy	27 (6.7)	250 (9.5)		2 (3.8)	117 (7.8)	
Conservative care	3 (0.7)	15 (0.6)		0 (0.0)	6 (0.4)	

Adapted from Kim et al. [19]
Bold indicates statistical significance
EBV Epstein Barr virus, *MSI* microsatellite instability, *MSS* microsatellite stable, *MSI-L* MSI low, *MSI-H* MSI-high

responses: for example, 17β-estradiol (E2) promotes IL-12 and interferon-γ production by dendritic cells, enhances B lymphocyte survival, triggers polyclonal B cell activation, and increases gut permeability, thereby fostering a pro-inflammatory milieu. Conversely, testosterone suppresses T cell proliferation and interferes with toll-like receptor (TLR)-mediated signaling pathways, distinguishing its effects from those of E2 on B cells.

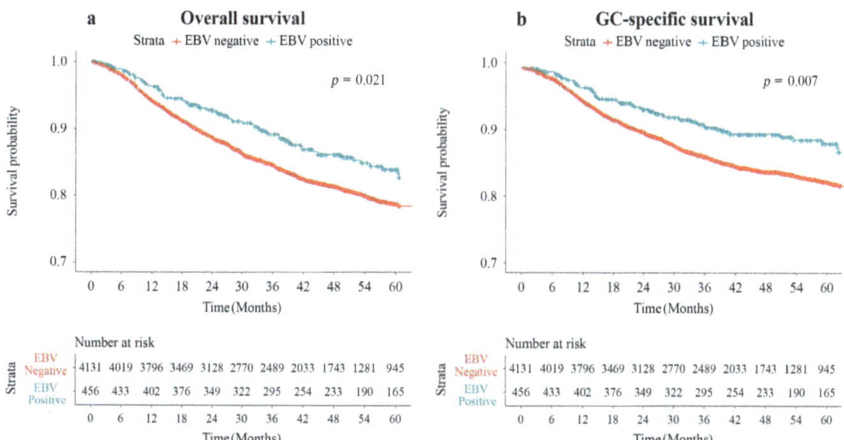

Fig. 8 Comparisons of overall survival and gastric cancer (GC)-specific survival according to Epstein–Barr virus (EBV) status. EBV positivity was significantly associated with improved outcomes for both overall survival (**a**) and GC-specific survival (**b**) (adapted from Kim et al. [19])

In our cohort, EBV positivity significantly improved prognosis in males, but not in females. This sex difference may be due to inherently stronger immune responses in females across various contexts, including cancer, viral infections, and microbial exposure. Prior studies have shown more extensive immune activation in GC among females compared to males [144, 145]. Furthermore, a study on tumor-associated neutrophils (TANs), a component of the tumor immune microenvironment, found that increased TAN infiltration was associated with improved prognosis in female GC patients only [146].

Although the exact biological mechanisms remain to be clarified, our findings suggest that the favorable prognostic impact of EBV positivity in males may stem from their relatively less activated baseline immune response, allowing the immunogenicity of EBV-associated tumors to exert a more pronounced survival benefit. In contrast, the stronger baseline immune activation in females may obscure this effect. Moreover, the male predominance of EBVaGC (4.03 times higher than in females) supports the hypothesis that sex-specific immune mechanisms contribute to its development and clinical behavior. Despite EBVaGC being frequently located proximally and exhibiting poorly differentiated histology, it conferred a survival advantage exclusively in males, further suggesting that EBV-related tumorigenesis may operate through sex-differentiated immune pathways.

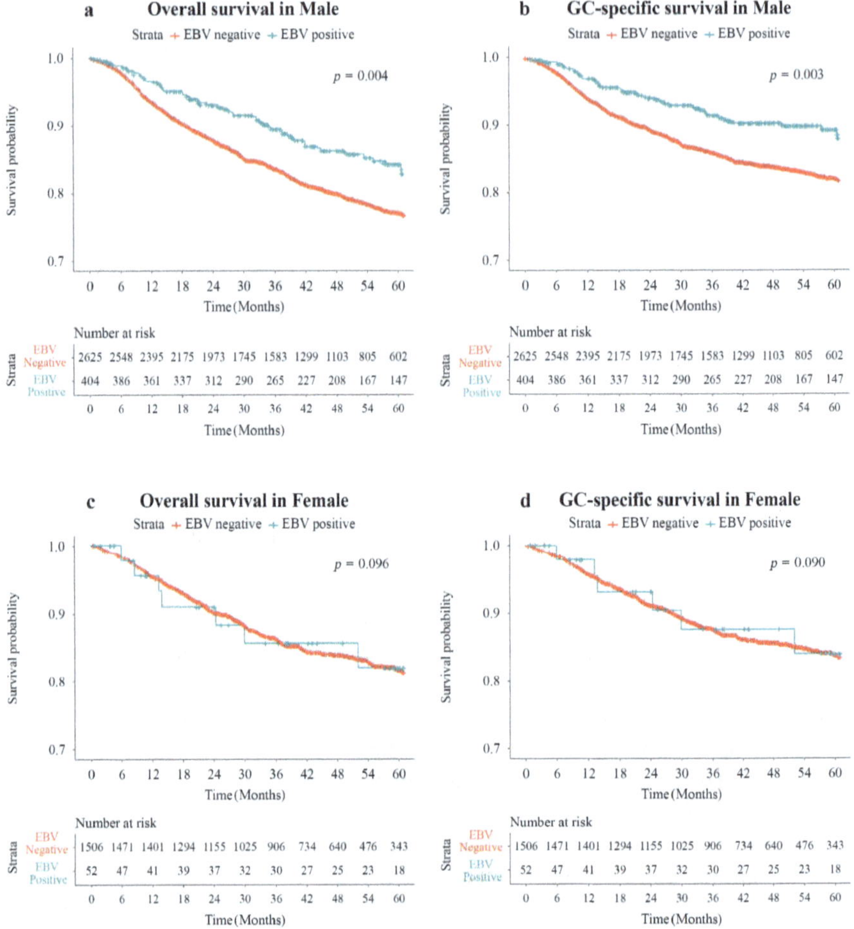

Fig. 9 Comparisons of overall survival and gastric cancer (GC)- specific survival according to Epstein-Barr virus (EBV) status in male and female. EBV positivity showed a positive effect significantly on both (**a**) overall and (**b**) GC-specific survival in male but no significant difference in female neither (**c**) overall nor (**d**) GC-specific survival (adapted from Kim et al. [19])

6 PD-L1 Expression and the Effect of Immune Checkpoint Inhibitors on Survival Outcomes in Patients with Gastric Cancer Depending on Sex

Sex plays a critical role in shaping immune system development and function. On average, females display more robust innate and adaptive immune responses compared to males [144]. Notable differences in T-cell polarization have been reported between the sexes, with males generally favoring a helper T cell type 1 (Th1) response, while females are more inclined toward a Th2 profile. Moreover,

Table 2 Univariate and multivariate analyses of risk factors for GC-specific survival according to sex

Variable		Male				Female			
		Univariate analyses HR [95% CI]	*p*-value	Multivariate analyses HR [95% CI]	*p*-value	Univariate analyses HR [95% CI]	*p*-value	Multivariate analyses HR [95% CI]	*p*-value
Age		1.032 (1.023–1.041)	<0.001	1.042 (1.031–1.054)	<0.001	1.000 (0.989–1.010)	0.932		
Smoking	Non-smoker	Ref		Ref		Ref			
	Smoker	0.747 (0.609–0.917)	**0.005**	0.996 (0.772–1.283)	0.973	1.531 (0.991–2.365)	0.055		
Size		1.297 (1.266–1.328)	**<0.001**	1.102 (1.063–1.142)	**<0.001**	1.290 (1.249–1.333)	**<0.001**	1.134 (1.079–1.191)	**<0.001**
Location	Upper	Ref		Ref		Ref		Ref	
	Middle	1.141 (0.887–1.467)	0.304	1.129 (0.817–1.560)	0.463	0.637 (0.460–0.883)	**0.007**	1.017 (0.614–1.683)	0.949
	Low	0.578 (0.460–0.728)	**<0.001**	0.847 (0.642–1.118)	0.242	0.242 (0.171–0.343)	**<0.001**	0.754 (0.481–1.182)	0.218
Histology	Differentiated	Ref		Ref		Ref		Ref	
	Undifferentiated	2.869 (2.319–3.549)	**<0.001**	1.260 (0.847–1.876)	0.254	1.992 (1.411–2.811)	**<0.001**	0.685 (0.431–1.088)	0.109
	Mixed	0.637 (0.389–1.044)	0.073	0.671 (0.368–1.222)	0.192	0.567 (0.314–1.022)	0.059	0.377 (0.195–0.730)	**0.004**
	Others	1.879 (1.249–2.827)	**0.002**	1.268 (0.752–2.139)	0.373	0.745 (0.231–2.404)	0.622	0.377 (0.090–1.586)	0.183

(continued)

Table 2 (continued)

Variable		Male				Female			
		Univariate analyses HR [95% CI]	*p*-value	Multivariate analyses HR [95% CI]	*p*-value	Univariate analyses HR [95% CI]	*p*-value	Multivariate analyses HR [95% CI]	*p*-value
Lauren	Intestinal	Ref		Ref		Ref			
	Diffuse	1.791 (1.415–2.267)	**<0.001**	1.172 (0.812–1.692)	0.398	1.396 (0.968–2.013)	0.074		
	Mixed	1.606 (0.819–3.150)	0.168	1.684 (0.775–3.657)	0.188	0.603 (0.187–1.945)	0.397		
	Indeterminate	4.794 (2.440–9.419)	**<0.001**	1.648 (0.737–3.684)	0.224	4.819 (0.662–35.085)	0.121		
TNM staging	Stage 1	Ref		Ref		Ref		Ref	
	Stage 2	7.425 (4.816–11.448)	**<0.001**	4.791 (3.045–7.539)	**<0.001**	6.317 (3.337–11.957)	**<0.001**	4.005 (2.050–7.827)	**<0.001**
	Stage 3	25.175 (17.304–36.627)	**<0.001**	13.084 (8.564–19.989)	**<0.001**	23.507 (13.722–40.272)	**<0.001**	11.817 (6.447–21.659)	**<0.001**
	Stage 4	103.874 (70.561–152.914)	**<0.001**	38.129 (20.594–70.595)	**<0.001**	120.019 (69.874–206.149)	**<0.001**	58.179 (28.131–120.322)	**<0.001**
EBV-ISH positivity	Negative	Ref		Ref		Ref			
	Positive	0.599 (0.428–0.840)	**0.003**	0.517 (0.327–0.816)	**0.005**	0.949 (0.421–2.138)	0.949		

Adapted from Kim et al. [19]
Bold indicates statistical significance
HR Hazard ratio, *EBV-ISH* Epstein–Barr virus in situ hybridization

males exhibit a greater skewing toward Th17 responses, whereas females demonstrate stronger B cell-mediated immunity, responding more effectively to various antigens and vaccines [147, 148].

There are also sex-based differences in immune evasion strategies. For example, in non-small cell lung cancer (NSCLC), the tumor immune microenvironment (TIME) in females is characterized by a higher infiltration of both innate and adaptive immune cells, including diverse T-cell subsets, compared to males. In contrast, male TIME more often demonstrates a T-cell-exclusion pattern [149].

At the hormonal level, sex hormones modulate the expression and functionality of immune checkpoints such as programmed cell death protein 1 (PD-1) and its ligand PD-L1. These hormonal effects play a key role in autoimmune susceptibility and immune modulation [150, 151]. Testosterone and estradiol can influence both innate and adaptive immune responses through their respective receptors. AR signaling has been shown to induce T cell exhaustion in CD^{8+} T cells by downregulating tumor necrosis factor (TNF), interferon gamma (IFN-γ), and granzyme B, thereby impairing antitumor immunity [152, 153].

In contrast, 17β-estradiol enhances ERβ expression, which is associated with reduced inflammation and increased expression of antioxidant enzymes, contributing to a protective immune profile in females [154]. However, ERα promotes polarization of tumor-associated macrophages (TAMs) toward the immunosuppressive M2 phenotype, reducing the antitumoral M1 population. This polarization contributes to CD^{8+} T cell dysfunction and can promote resistance to immune checkpoint inhibitors (ICIs) targeting the PD-1/PD-L1 pathway [69, 155].

Taken together, these findings suggest that the expression and therapeutic targeting of PD-L1, as well as the response to PD-1/PD-L1 blockade in GC, may differ significantly between males and females due to intrinsic hormonal and immune regulatory mechanisms.

6.1 PD-L1 and Immune Checkpoint Inhibitors Targeting PD-1/PD-L1

PD-L1, a transmembrane glycoprotein belonging to the B7 superfamily [156, 157], is expressed in various cell types, including immune cells, dendritic cells, epithelial cells, and certain tumor cells. Members of the B7 family modulate immune responses by suppressing T-cell activity through engagement with the programmed death-1 (PD-1) receptor [158, 159]. The PD-1/PD-L1 interaction serves as a critical immune checkpoint to prevent hyperactivation of adaptive immunity and autoimmunity [160].

In the context of cancer, PD-L1 acts as an immune suppressor by interacting with PD-1, which is abundantly expressed on tumor-specific T cells. This interaction inhibits T-cell receptor-mediated signaling and impairs both innate and adaptive immunity, particularly by suppressing the cytotoxic function of CD^{8+} T cells [161–163]. ICIs targeting the PD-1/PD-L1 axis have demonstrated clinical efficacy in

GC and other malignancies. However, therapeutic responses vary among patients, reflecting the biological heterogeneity of tumors.

These variations are influenced not only by intrinsic tumor properties such as genetic mutations, histological features, and pTNM stage but also by extrinsic factors, including the TIME. Chronic infections, such as *H. pylori* and EBV, as well as microsatellite instability (MSI), can trigger inflammation that contributes to the formation of premalignant lesions. Tumors also evade immune surveillance by altering the function of immune cells within the TIME.

Notably, tumor-infiltrating lymphocytes (TILs)—particularly CD^{8+} cytotoxic T cells that are critical for tumor clearance—can become dysfunctional after sustained exposure to the tumor microenvironment. Thus, the TIME is central to GC pathogenesis and progression [129, 164]. Because of this, evaluating the characteristics of the TIME, including the infiltration and functionality of immune cells, is essential for predicting responses to immunotherapies. PD-L1 is not only a therapeutic target but also a surrogate marker for TIL density and immune activity within tumors [165].

Sex-based immune differences further influence PD-L1 dynamics. Males tend to mount more pronounced inflammatory responses, while females often exhibit relatively suppressed inflammation, which may lead to enhanced PD-1 expression. Interestingly, in several cohorts of patients with advanced melanoma, a higher body mass index (BMI) was associated with improved responses to ICIs—but only in males [166], despite obesity being a known cancer risk factor due to persistent inflammation [166]. These observations suggest that PD-L1 expression and immune checkpoint efficacy in GC may be modulated by sex-dependent differences in immune regulation.

6.2 *Clinicopathological Factors Influencing PD-L1 Expression and the Effect of Immune Checkpoint Inhibitors on Survival Outcomes in Patients with Gastric Cancer Depending on Sex*

Our team analyzed the determinants of PD-L1 expression in GC tissues by examining 470 Korean patients, focusing on clinicopathological features and prognostic relevance with particular attention to sex differences [20]. Among these, 281 individuals (59.7%) were PD-L1 positive, comprising 189 males and 92 females. In the overall cohort as well as in male patients, EBV infection was strongly associated with PD-L1 positivity (overall: odds ratio [OR] 19.18, $p = 0.005$; males: OR 14.55, $p = 0.01$), as was tumor location in the antrum (overall: OR 1.34, $p = 0.03$; males: OR 1.38, $p = 0.01$). In contrast, among females, intestinal-type histology exhibited a marginally inverse association with PD-L1 expression (OR 0.57, $p = 0.05$) [20].

Regarding treatment response, the administration of ICIs did not significantly improve overall survival in the entire cohort (Fig. 10a) [20]. However, sex-stratified analysis revealed that male patients receiving ICI-containing chemotherapy had significantly prolonged overall survival compared to those not receiving ICIs (*p*

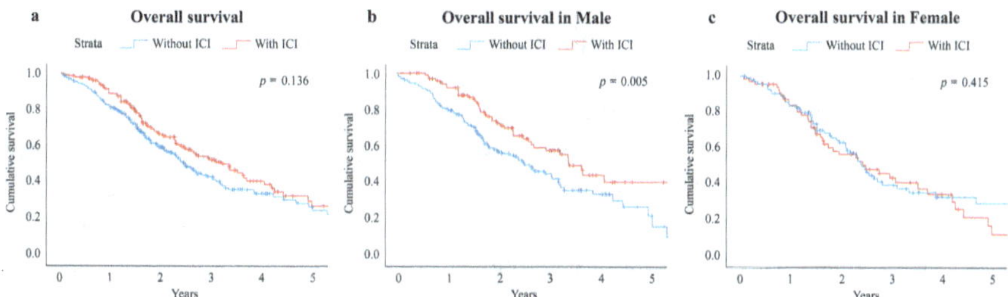

Fig. 10 Comparisons of cumulative survival according to immune checkpoint inhibitor (ICI) treatment by sex. (**a**) In the overall cohort, no significant survival difference was observed between ICI-treated and non-ICI-treated groups. (**b**) Among male patients, ICI treatment was associated with significantly improved survival. (**c**) In female patients, no significant survival difference was observed based on ICI treatment (adapted from Lee et al. [20])

< 0.05) (Fig. 10b). In contrast, no survival benefit was observed in female patients (Fig. 10c) [20].

These findings indicate that EBV infection and antral tumor location are independent predictors of PD-L1 expression specifically in male GC patients. Additionally, the survival advantage conferred by ICI therapy appears to be limited to males, underscoring the potential need for sex-specific approaches in immunotherapy for GC [20].

The prognostic value of PD-L1 expression in GC remains inconclusive, with prior studies reporting positive, negative, or no significant associations with clinical outcomes [167]. In our analysis, PD-L1 expression, defined as a Combined Positive Score (CPS) ≥ 5, did not correlate significantly with survival in either the total cohort or in sex-stratified subgroups. Therefore, PD-L1 positivity alone did not function as an independent prognostic marker in GC patients [20]. Notably, Choi et al. proposed that the combined evaluation of PD-L1 status and CD^{8+} tumor-infiltrating lymphocytes (TILs) could serve as a more reliable prognostic indicator [168]. Their findings indicated that poor prognosis was associated with tumors that were PD-L1-positive in tumor cells but negative in immune cells, alongside low CD^{8+} TIL infiltration. Supporting this, Noh et al. demonstrated that tumors exhibiting both PD-L1 positivity and high CD^{8+} TIL levels had the most favorable outcomes, while PD-L1-positive tumors with low CD^{8+} TIL density had the poorest prognosis [169]. These results highlight that high TIL density is generally linked to better prognosis and that PD-L1 expression is strongly associated with TIL infiltration. The favorable prognosis observed in EBVaGC, often accompanied by high PD-L1 levels, may also be partially attributed to dense TIL presence [165, 170].

While numerous studies have attempted to determine whether PD-L1 expression predicts GC outcomes, few have specifically considered sex as a modifying factor—despite well-documented immune system differences between sexes. Clinical responses to ICI therapy in cancers such as melanoma and non-small cell lung cancer (NSCLC) have been shown to vary by sex [171]. In our cohort, male GC patients who received ICI-based regimens experienced a significant improvement in survival (Fig. 10a), an effect not observed in female counterparts (Fig. 10b) [20].

This disparity may arise from sex-based differences in immunobiology. In males, androgen-mediated immunosuppression leads to weakened T-cell surveillance and the persistence of immunogenic cancer cells. These exhausted T-cells, often expressing high levels of PD-L1 or CTLA-4, are more susceptible to reinvigoration by ICIs, resulting in stronger therapeutic responses [149, 172, 173]. Conversely, females typically exhibit heightened innate and adaptive immune responses, including more efficient activation and differentiation of CD^{8+} T-cells [174]. Hormonal modulators like estrogen and selective ER modulators (SERMs) have been shown to influence key immune checkpoints such as PD-1, PD-L1, and CTLA-4, functioning in ways that may parallel ICI mechanisms [175].

Castro et al. posited that early-stage tumors in females may be more aggressively eliminated due to their robust immune environments, forcing surviving tumors to adapt through immune evasion. These tumors accumulate mutations that reduce antigen presentation via MHC-I and MHC-II pathways, leading to lower immunogenicity and greater resistance in advanced disease stages [176, 177]. Consequently, advanced GC in females may harbor more sophisticated immune escape mechanisms than in males, potentially explaining the diminished benefit from ICIs observed in female patients [177]. Furthermore, the TIME in females tends to be richer in immune cells but also in immunosuppressive elements, including higher expression of alternative checkpoints (e.g., TIM3, TIGIT, VISTA) and the presence of suppressive cell populations such as cancer-associated fibroblasts (CAFs) and myeloid-derived suppressor cells [171]. Despite more efficient early immune responses, the accumulation of redundant escape mechanisms may lead to profound T-cell exhaustion in females.

In summary, our study underscores that PD-L1 expression and ICI efficacy in GC differ significantly between sexes. Male patients, unlike females, demonstrated clear survival benefits from ICI-containing therapies, highlighting the critical need for sex-tailored treatment strategies in GC management [20].

7 Conclusions

GC is generally more prevalent in males than females, with a reported male-to-female ratio of approximately 2:1. However, in patients younger than 40 years, the trend reverses, with 67.3% of cases occurring in females. This contrasts with the pattern observed in those aged over 50, where 65–70% of cases are male. Among female patients, the diffuse-type histology predominates before menopause, while the frequency of the intestinal-type gradually aligns with that in males about two decades post-menopause. Four types of sex hormone receptors—ERα, ERβ, progesterone receptor, and AR—are independently expressed in GC tissues, suggesting a partial role of sex hormones in gastric tumorigenesis [11]. To date, ERα and AR have been positively associated with tumor development, whereas ERβ expression has been inversely correlated. ERα is linked with diffuse or mixed histology and poorly differentiated tumors, while AR is more commonly observed in cardia cancer and undifferentiated tumors.

Furthermore, the influence of overweight and obesity on GC risk differs by sex. Recent studies also highlight a rising concern regarding the increased risk of non-cardia GC among underweight individuals, with potential sex-related variations in this association. The heightened risk of cardia GC in the context of overweight/obesity may be modulated by obesity-induced elevations in female hormone levels and sex-specific susceptibility to precancerous lesions in the cardia. Conversely, emerging evidence suggests that underweight status is more strongly associated with non-cardia GC, in contrast to obesity's link with cardia GC.

In terms of viral oncogenesis, sex differences have also been noted in EBVaGC. In both sexes, EBV-positive tumors were primarily located in the upper third of the stomach. However, in males, EBV positivity was significantly associated with diffuse-type histology, a pattern not observed in females. Notably, EBVaGC patients showed a better prognosis, which was limited to male patients, implying that EBV-related gastric carcinogenesis may be influenced by sex-specific immune pathways. In addition, EBV positivity and antral tumor location were independently associated with PD-L1 expression in male, but not female, patients with GC [20]. Moreover, ICI-based therapy significantly improved survival in male patients, while no such benefit was seen in females [20].

In conclusion, GC demonstrates distinct sex- and gender-related biological characteristics. These differences underscore the importance of designing and conducting future research and treatment strategies with sex-specific considerations in mind.

Acknowledgements This work was supported by the National Institute of Health (NIH) research project (project No. 2025-ER1104-00).

References

1. Joo YE, Park HK, Myung DS, Baik GH, Shin JE, Seo GS, et al. Prevalence and risk factors of atrophic gastritis and intestinal metaplasia: a nationwide multicenter prospective study in Korea. Gut Liver. 2013;7:303–10.
2. Smyth EC, Nilsson M, Grabsch HI, van Grieken NC, Lordick F. Gastric cancer. Lancet. 2020;396:635–48.
3. Kim N. Sex/gender-specific medicine in the gastrointestinal disease. Singapore: Springer; 2022.
4. Liu Z, Zhang Y, Lagergren J, Li S, Li J, Zhou Z, et al. Circulating sex hormone levels and risk of gastrointestinal cancer: systematic review and meta-analysis of prospective studies. Cancer Epidemiol Biomarkers Prev. 2023;32:936–46.
5. Lee S, Kim KM, Lee SY, Jung J. Estrogen aggravates tumor growth in a diffuse gastric cancer xenograft model. Pathol Oncol Res. 2021;27:622733.
6. Ryu WS, Kim JH, Jang YJ, Park SS, Um JW, Park SH, et al. Expression of estrogen receptors in gastric cancer and their clinical significance. J Surg Oncol. 2012;106:456–61.

7. Tang W, Liu R, Yan Y, Pan X, Wang M, Han X, et al. Expression of estrogen receptors and androgen receptor and their clinical significance in gastric cancer. Oncotarget. 2017;8:40765–77.
8. Wang X, Xia X, Xu E, Yang Z, Shen X, Du S, et al. Estrogen receptor beta prevents signet ring cell gastric carcinoma progression in young patients by inhibiting pseudopodia formation via the mTOR–Arpc1b/EVL signaling pathway. Front Cell Dev Biol. 2021;8:592919.
9. Ur Rahman MS, Cao J. Estrogen receptors in gastric cancer: advances and perspectives. World J Gastroenterol. 2016;22:2475–82.
10. Ge H, Yan Y, Tian F, Wu D, Huang Y. Prognostic value of estrogen receptor α and estrogen receptor β in gastric cancer based on a meta-analysis and The Cancer Genome Atlas (TCGA) datasets. Int J Surg. 2018;53:24–31.
11. Gan L, He J, Zhang X, Zhang YJ, Yu GZ, Chen Y, et al. Expression profile and prognostic role of sex hormone receptors in gastric cancer. BMC Cancer. 2012;12:566.
12. Cancer Genome Atlas Research Network. Comprehensive molecular characterization of gastric adenocarcinoma. Nature. 2014;513:202–9.
13. Chia NY, Tan P. Molecular classification of gastric cancer. Ann Oncol. 2016;27:763–9.
14. IARC Working Group on the Evaluation of Carcinogenic Risks to Humans. Epstein-Barr virus and Kaposi's sarcoma herpesvirus/human herpesvirus 8, Lyon: International Agency for Research on Cancer; 1997. p. 47–262.
15. Tsao SW, Tsang CM, To KF, Lo KW. The role of Epstein-Barr virus in epithelial malignancies. J Pathol. 2015;235:323–33.
16. Cristescu R, Lee J, Nebozhyn M, Kim KM, Ting JC, Wong SS, et al. Molecular analysis of gastric cancer identifies subtypes associated with distinct clinical outcomes. Nat Med. 2015;21:449–56.
17. Naseem M, Barzi A, Brezden-Masley C, Puccini A, Berger MD, Tokunaga R, et al. Outlooks on Epstein-Barr virus associated gastric cancer. Cancer Treat Rev. 2018;66:15–22.
18. Kim YS, Nam SC, Han MH, Jeong JY, Park SK, Suh IS, et al. Predictive factors of Epstein-Barr virus association in gastric adenocarcinoma. Korean J Pathol. 2008;42:193–7.
19. Kim JH, Kim N, Song DH, Choi Y, Jeon EB, Kim S, et al. Sex-dependent different clinicopathological characterization of Epstein-Barr virus associated gastric carcinoma: a large-scale study. Gastric Cancer. 2024;27:221–34.
20. Lee JH, Kim N, Kim JH, Oh HJ, Kim Y, Choi Y, et al. Clinicopathological factors influencing PD-L1 expression and the effect of immune checkpoint inhibitors on survival outcomes in patients with gastric cancer depending on sex in a tertiary hospital in South Korea. Cancer Res Ther. 2025. https://doi.org/10.4143/crt.2025.126.
21. Ferlay J, Bray F, Steliarova-Foucher E, Forman D. Cancer incidence in five continents, CI5plus. IARC CancerBase No. 9. Lyon: International Agency for Research on Cancer; 2014.
22. Jemal A, Bray F, Center MM, Ferlay J, Ward E, Forman D, et al. Global cancer statistics. CA Cancer J Clin. 2011;61:69–90.
23. Sipponen P, Correa P. Delayed rise in incidence of gastric cancer in females results in unique sex ratio (M/F) pattern: etiologic hypothesis. Gastric Cancer. 2002;5:213–9.
24. Ferlay J, Soerjomataram I, Dikshit R, Eser S, Mathers C, Rebelo M, et al. Cancer incidence and mortality worldwide: sources, methods and major patterns in GLOBOCAN 2012. Int J Cancer. 2015;136:E359–86.
25. Kolonel LN, Hankin JH, Nomura AMY. Multiethnic studies of diet, nutrition, and cancer in Hawaii. In: Hayashi Y, Nagao M, Sugimura T, Tokoyama S, Tomatis L, Wattenberg LW, et al, editors. Diet, nutrition and cancer. Proceedings of the 16th international symposium of the princess Takamatsu cancer research fund. Utrecht: Tokyo/VNU Science; 1986. p. 29–40.
26. IARC Working Group on the Evaluation of Carcinogenic Risks to Humans. Infection with *Helicobacter pylori*. IARC Monogr Eval Carcinog Risks Hum. 1994;61:177–240.
27. Plummer M, Franceschi S, Vignat J, Forman D, de Martel C. Global burden of gastric cancer attributable to *Helicobacter pylori*. Int J Cancer. 2015;136:487–90.
28. World Cancer Research Fund/American Institute for Cancer Research (WCRF/AICR). Continuous update project report: diet, nutrition, physical activity and stomach cancer 2016.

Revised 2018. London: World Cancer Research Fund International; 2018. https://www.wcrf.org/sites/default/files/Stomach-Cancer-2016-Report.pdf. Accessed 21 June 2018.
29. IARC Working Group on the Evaluation of Carcinogenic Risks to Humans. Personal habits and indoor combustions. Volume 100 E. A review of human carcinogens. IARC Monogr Eval Carcinog Risks Hum. 2012;100 pt E:1–538.
30. Bray F, Ferlay J, Soerjomataram I, Siegel RL, Torre LA, Jemal A. Global cancer statistics 2018: GLOBOCAN estimates of incidence and mortality worldwide for 36 cancers in 185 countries. Cancer J Clin. 2018;68:394–424.
31. Howson CP, Hiyama T, Wynder EL. The decline in gastric cancer: epidemiology of an unplanned triumph. Epidemiol Rev. 1986;8:1–27.
32. deMartel C, Parsonnet J. Stomach cancer. In: Thun MJ, Linet MS, Cerhan JR, Haiman CA, Schottenfeld D, editors. Cancer epidemiology and prevention. 4th ed. New York: Oxford University Press; 2018. p. 593–610.
33. Takekura N, Takanashi A, Tabuchi J, Haruta R, Tahara E. Estrogen receptors in gastric adenocarcinoma: a retrospective immunohistochemical analysis. Virchows Arch A Pathol Anat Histopathol. 1988;413:297–302.
34. Matsui M, Kojima O, Uehara Y, Takahashi T. Characterization of estrogen receptor in human gastric cancer. Cancer. 1991;68:305–8.
35. Singh S, Poulsom R, Wright NA, Sheppard MC, Langman MJ. Differential expression of oestrogen receptor and oestrogen inducible genes in gastric mucosa and cancer. Gut. 1997;40:516–20.
36. Yi JH, Do IG, Jang J, Kim ST, Kim KM, Park SH, et al. Anti-tumor efficacy of fulvestrant in estrogen receptor positive gastric cancer. Sci Rep. 2014;4:7592.
37. Kominea A, Konstantinopoulos PA, Kapranos N, Vandoros G, Gkermpesi M, Andricopoulos P, et al. Androgen receptor (AR) expression is an independent unfavorable prognostic factor in gastric cancer. J Cancer Res Clin Oncol. 2004;130:253–8.
38. Baek SM, Kim N, Kwon YJ, Lee HS, Kim HY, Lee J, et al. Serum pepsinogen II and *Helicobacter pylori* status in the detection of diffuse type of early gastric cancer at young age in South Korea. Gut Liver. 2020;14:439–49.
39. Tian Y, Wan H, Lin Y, Xie X, Li Z, Tan G. Androgen receptor may be responsible for gender disparity in gastric cancer. Med Hypotheses. 2013;80:672–4.
40. Wu CW, Chi CW, Chang TJ, Lui WY, P'eng FK. Sex hormone receptors in gastric cancer. Cancer. 1990;65:1396–400.
41. Nakamura Y, Shimada N, Suzuki T, Imatani A, Sekine H, Ohara S, et al. In situ androgen production in human gastric carcinoma–androgen synthesizing and metabolizing enzymes. Anticancer Res. 2006;26:1935–9.
42. Song M, Rabkin CS, Camargo MC. Gastric cancer: an evolving disease. Curr Treat Options Gastroenterol. 2018;16:561–9.
43. Kim SM, Min BH, Lee J, An JY, Lee JH, Sohn TS, et al. Protective effects of female reproductive factors on Lauren intestinal-type gastric adenocarcinoma. Yonsei Med J. 2018;59:28–34.
44. Jung YJ, Kim HJ, Park CH, Park SJ, Kim N. Effects of female reproductive factors on Lauren intestinal-type gastric cancer in female; A multicenter retrospective study in South Korea. Gut Liver. 2022;16:706–15.
45. Deng H, Huang X, Fan J, Wang L, Xia Q, Yang X, et al. A variant of estrogen receptor-alpha, ER-alpha36 is expressed in human gastric cancer and is highly correlated with lymph node metastasis. Oncol Rep. 2010;24:171–6.
46. Choi Y, Kim N, Jo HH, Park J, Yoon H, Shin CM, et al. Sex differences in histology, staging, and prognosis in 2,983 patients who received gastric cancer surgery. World J Gastroenterol. 2022;28:933–47.
47. Sica V, Nola E, Contieri E, Bova R, Masucci MT, Medici N, et al. Estradiol and progesterone receptors in malignant gastrointestinal tumors. Cancer Res. 1984;44:4670–4.
48. Xu CY, Guo JL, Jiang ZN, Xie SD, Shen JG, Shen JY, et al. Prognostic role of estrogen receptor alpha and estrogen receptor beta in gastric cancer. Ann Surg Oncol. 2010;17:2503–9.

49. Guo JL, Xu CY, Jiang ZN, Dong MJ, Xie SD, Shen JG, et al. Estrogen receptor beta variants mRNA expressions in gastric cancer tissues and association with clinicopathologic parameters. Hepatogastroenterology. 2009;57:1584–8.
50. Lee HR, Hwang KA, Park MA, Yi BR, Jeung EB, Choi KC, et al. Treatment with bisphenol A and methoxychlor results in the growth of human breast cancer cells and alteration of the expression of cell cycle-related genes, cyclin D1 and p21, via an estrogen receptor-dependent signaling pathway. Int J Mol Med. 2012;29:883–90.
51. Park MA, Hwang KA, Choi KC. Diverse animal models to examine potential role(s) and mechanism of endocrine disrupting chemicals on the tumor progression and prevention: Do they have tumorigenic or anti-tumorigenic property? Lab Anim Res. 2011;27:265–73.
52. Chen P, Li B, OuYang L. Role of estrogen receptos in health and disease. Front Endocrinol (Lausanne). 2022;13:839005.
53. Tan MH, Li J, Xu He, Melcher K, Yong EL. Androgen receptor: structure, role in prostate cancer and drug discovery. Acta Pharmacol Sin. 2015;36:3–23.
54. Matsuyama S, Ohkura Y, Eguchi H, Kobayashi Y, Akagi K, Uchida K, et al. Estrogen receptor beta is expressed in human stomach adenocarcinoma. J Cancer Res Clin Oncol. 2002;128:319–24.
55. Choi Y, Kim N, Park JH, Song CH, Oh HJ. Expression rates of sex hormone receptors with their clinical Correlates in gastric cancer patients and normal controls. World J Mens Health. 2025;43:980–91.
56. Freedman ND, Chow WH, Gao YT, Shu XO, Ji BT, Yang G, et al. Menstrual and reproductive factors and gastric cancer risk in a large prospective study of women. Gut. 2007;56:1671–7.
57. Kim HW, Kim JH, Lim BJ, Kim H, Kim H, Park JJ, et al. Sex disparity in gastric cancer: female sex is a poor prognostic factor for advanced gastric cancer. Ann Surg Oncol. 2016;23:4344–51.
58. Li H, Wei Z, Wang C, Chen W, He Y, Zhang C. Gender differences in gastric cancer survival: 99,922 cases based on the SEER database. J Gastrointest Surg. 2020;24:1747–57.
59. Jaspers VK, Gillessen A, Quakernack K. Gastric cancer in pregnancy: do pregnancy, age or female sex alter the prognosis? Case reports and review. Eur J Obstet Gynecol Reprod Biol. 1999;87:13–22.
60. Kang S, Park M, Cho JY, Ahn SJ, Yoon C, Kim SK, Cho SJ. Tumorigenic mechanisms of estrogen and *Helicobacter pylori* cytotoxin-associated gene A in estrogen receptor α-positive diffuse-type gastric adenocarcinoma. Gastric Cancer. 2022;25:678–96.
61. Wang M, Pan JY, Song GR, Chen HB, An LJ, Qu SX. Altered expression of estrogen receptor alpha and beta in advanced gastric adenocarcinoma: correlation with prothymosin alpha and clinicopathological parameters. Eur J Surg Oncol. 2007;33:195–201.
62. Wang X, Deng H, Zou F, et al. ER-α36-mediated gastric cancer cell proliferation via the c-Src pathway. Oncol Lett. 2013;6:329–35.
63. Fu Z, Deng H, Wang X, et al. Involvement of ER-α36 in the malignant growth of gastric carcinoma cells is associated with GRP94 overexpression. Histopathology. 2013;63:325–33.
64. Williams C, DiLeo A, Niv Y, Gustafsson JÅ. Estrogen receptor beta as target for colorectal cancer prevention. Cancer Lett. 2016;372:48–56.
65. Indukuri R, Jafferali MH, Song D, et al. Genome-wide estrogen receptor β chromatin binding in human colon cancer cells reveals its tumor suppressor activity. Int J Cancer. 2021;149:692–706.
66. Jukic Z, Radulovic P, Stojković R, Mijic A, Grah J, Kruslin B, et al. Gender difference in distribution of estrogen and androgen receptors in intestinal-type gastric cancer. Anticancer Res. 2017;37:197–202.
67. Hsu LW, Huang KH, Chen MH, Fang WL, Chao Y, Lo SS, et al. Genetic alterations in gastric cancer patients according to sex. Aging (Albany NY). 2020;13:376–88.
68. Böger C, Behrens HM, Mathiak M, Krüger S, Kalthoff H, Röcken C. PD-L1 is an independent prognostic predictor in gastric cancer of western patients. Oncotarget. 2016;7:24269–83.
69. Özdemir BC, Dotto GP. Sex hormones and anticancer immunity. Clin Cancer Res. 2019;25:4603–10.

70. Basen-Engquist K, Chang M. Obesity and cancer risk: recent review and evidence. Curr Oncol Rep. 2011;13:71–6.
71. World Cancer Research Fund/American Institute for Cancer Research. Food, nutrition, physical activity, and the prevention of cancer: a global perspective. Washington DC: AICR; 2007.
72. Renehan AG, Tyson M, Egger M, Heller RF, Zwahlen M. Body-mass index and incidence of cancer: a systematic review and meta-analysis of prospective observational studies. Lancet. 2008;371:569–78.
73. Mukaisho KI, Nakayama T, Hagiwara T, Hattori T, Sugihara H. Two distinct etiologies of gastric cardia adenocarcinoma: interactions among pH, *Helicobacter pylori*, and bile acids. Front Microbiol. 2015;6:412.
74. Fan JH, Wang JB, Wang SM, Abnet CC, Qiao YL, Taylor PR. Body mass index and risk of gastric cancer: a 30-year follow-up study in the Linxian general population trial cohort. Cancer Sci. 2017;108:1667–72.
75. Hirabayashi M, Inoue M, Sawada N, Saito E, Abe SK, Hidaka A, et al. Effect of body-mass index on the risk of gastric cancer: a population-based cohort study in a Japanese population. Cancer Epidemiol. 2019;63:101622.
76. Jang J, Cho EJ, Hwang Y, Weiderpass E, Ahn C, Choi J, et al. Association between body mass index and gastric cancer risk according to effect modification by *Helicobacter pylori* infection. Cancer Res Treat. 2019;51:1107–16.
77. Jang J, Wang TY, Cai H, Ye F, Murphy G, Shimazu T, et al. The U-shaped association between body mass index and gastric cancer risk in the *Helicobacter pylori* Biomarker Cohort Consortium: a nested case-control study from eight East Asian cohort studies. Int J Cancer. 2020;147:777–84.
78. Chen Y, Liu LX, Wang XL, Wang J, Yan Z, Cheng J, et al. Body mass index and risk of gastric cancer: a meta-analysis of a population with more than ten million from 24 prospective studies. Cancer Epidem Biomar. 2013;22:1395–408.
79. NCD Risk Factor Collaboration. Worldwide trends in body-mass index, underweight, overweight, and obesity from 1975 to 2016: a pooled analysis of 2,416 population-based measurement studies in 128.9 million children, adolescents, and adults. Lancet. 2017;390:2627–42.
80. Guh DP, Zhang W, Bansback N, Amarsi Z, Birmingham CL, Anis AH. The incidence of co-morbidities related to obesity and overweight: a systematic review and meta-analysis. BMC Public Health. 2009;9:88.
81. Lauby-Secretan B, Scoccianti C, Loomis D, Grosse Y, Bianchini F, Straif K, et al. Body fatness and cancer—viewpoint of the IARC working group. New Engl J Med. 2016;375:794–8.
82. World Cancer Research Fund/American Institute for Cancer Research. Diet, nutrition, physical activity and oesophageal cancer. In: Continuous update project expert report 2018; 2018. dietandcancerreport.org. Accessed 18 Aug 2021.
83. Pearson-Stuttard J, Zhou B, Kontis V, Bentham J, Gunter MJ, Ezzati M. Worldwide burden of cancer attributable to diabetes and high body-mass index: a comparative risk assessment. Lancet Diabetes Endocrinol. 2018;6:95–104.
84. Jang J, Choi Y, Kim N. Sex differences in the effect of changes in body mass index on the risk of developing gastric cancer: findings from a nationwide retrospective cohort study. Gut Liver. 2025. https://doi.org/10.5009/gnl240555.
85. Derakhshan MH, Malekzadeh R, Watabe H, Yazdanbod A, Fyfe V, Kazemi A, et al. Combination of gastric atrophy, reflux symptoms and histological subtype indicates two distinct aetiologies of gastric cardia cancer. Gut. 2008;57:298–305.
86. Horii T, Koike T, Abe Y, Kikuchi R, Unakami H, Iijima K, et al. Two distinct types of cancer of different origin may be mixed in gastroesophageal junction adenocarcinomas in Japan: evidence from direct evaluation of gastric acid secretion. Scand J Gastroenterol. 2011;46:710–9.
87. Alemán JO, Eusebi LH, Ricciardiello L, Patidar K, Sanyal AJ, Holt PR. Mechanisms of obesity-induced gastrointestinal neoplasia. Gastroenterology. 2014;146:357–73.

88. Olefson S, Moss SF. Obesity and related risk factors in gastric cardia adenocarcinoma. Gastric Cancer. 2015;18:23–32.
89. Corley DA, Kubo A. Body mass index and gastroesophageal reflux disease: a systematic review and meta-analysis. Am J Gastroenterol. 2006;101:2619–28.
90. Weston AP, Krmpotich PT, Cherian R, Dixon A, Topalovski M. Prospective evaluation of intestinal metaplasia and dysplasia within the cardia of patients with Barrett's esophagus. Dig Dis Sci. 1997;42:597–602.
91. Cameron AJ, Lomboy CT, Pera M, Carpenter HA. Adenocarcinoma of the esophagogastric junction and Barrett's esophagus. Gastroenterology. 1995;109:1541–6.
92. Fullard M, Kang JY, Neild P, Poullis A, Maxwell JD. Systematic review: does gastro-oesophageal reflux disease progress? Aliment Pharmacol Ther. 2006;24:33–45.
93. Cohen DH, LeRoith D. Obesity, type 2 diabetes, and cancer: the insulin and IGF connection. Endocr Relat Cancer. 2012;19:F27-45.
94. Tseng CH, Chen CJ, Landolph JR Jr. Diabetes and cancer: epidemiological, clinical, and experimental perspectives. Exp Diabetes Res. 2012;2012:101802.
95. Kim HJ, Kim N, Kim HY, Lee HS, Yoon H, Shin CM, et al. Relationship between body mass index and the risk of early gastric cancer and dysplasia regardless of *Helicobacter pylori* infection. Gastric Cancer. 2015;18:762–73.
96. Camargo MC, Goto Y, Zabaleta J, Morgan DR, Correa P, Rabkin CS. Sex hormones, hormonal interventions, and gastric cancer risk: a meta-analysis. Cancer Epidemiol Biomarkers Prev. 2012;21:20–38.
97. Lukanova A, Lundin E, Zeleniuch-Jacquotte A, Muti P, Mure A, Rinaldi S, et al. Body mass index, circulating levels of sex-steroid hormones, IGF-I and IGF-binding protein-3: a cross-sectional study in healthy women. Eur J Endocrinol. 2004;150:161–71.
98. Madigan MP, Troisi R, Potischman N, Dorgan JF, Brinton LA, Hoover RN. Serum hormone levels in relation to reproductive and lifestyle factors in postmenopausal women (United States). Cancer Causes Control. 1998;9:199–207.
99. McTiernan A, Wu L, Chen C, Chlebowski R, Mossavar-Rahmani Y, Modugno F, et al. Relation of BMI and physical activity to sex hormones in postmenopausal women. Obesity (Silver Spring). 2006;14:1662–77.
100. Hankinson SE, Willett WC, Manson JE, Hunter DJ, Colditz GA, Stampfer MJ, et al. Alcohol, height, and adiposity in relation to estrogen and prolactin levels in postmenopausal women. J Natl Cancer Inst. 1995;87:1297–302.
101. Boyapati SM, Shu XO, Gao YT, Dai Q, Yu H, Cheng JR, et al. Correlation of blood sex steroid hormones with body size, body fat distribution, and other known risk factors for breast cancer in post-menopausal Chinese women. Cancer Causes Control. 2004;15:305–11.
102. Bezemer ID, Rinaldi S, Dossus L, Gils CH, Peeters PH, Noord PA, et al. C-peptide, IGF-I, sex-steroid hormones and adiposity: a cross-sectional study in healthy women within the European prospective investigation into cancer and nutrition (EPIC). Cancer Causes Control. 2005;16:561–72.
103. Choi IY, Choi YJ, Shin DW, Han KD, Jeon KH, Jeong S-M, et al. Association between obesity and the risk of gastric cancer in premenopausal and postmenopausal women: a nationwide cohort study. J Gastroenterol Hepatol. 2021;36:2834–40.
104. Ronkainen J, Talley NJ, Storskrubb T, Johansson SE, Lind T, Vieth M, et al. Erosive esophagitis is a risk factor for Barrett's esophagus: a community-based endoscopic follow-up study. Am J Gastroenterol. 2011;106:1946–52.
105. Kim N, Lee SW, Cho SI, Park CG, Yang CH, Kim HS, et al. The prevalence of and risk factors for erosive oesophagitis and non-erosive reflux disease: a nationwide multicentre prospective study in Korea. Aliment Pharmacol Ther. 2008;27:173–85.
106. Kim YS, Kim N, Kim GH. Sex and gender differences in gastroesophageal reflux disease. J Neurogastroenterol Motil. 2016;22:575–88.
107. Cossentino MJ, Wong RK. Barrett's esophagus and risk of esophageal adenocarcinoma. Semin Gastrointest Dis. 2003;14:128–35.

108. Kim SY, Jung HK, Lim J, Kim TO, Choe AR, Tae CH, et al. Gender specific differences in prevalence and risk factors for gastro-esophageal reflux disease. J Korean Med Sci. 2019;34:e158.
109. Grishina I, Fenton A, Sankaran-Walters S. Gender differences, aging and hormonal status in mucosal injury and repair. Aging Dis. 2014;52:160–9.
110. Kendall BJ, Macdonald GA, Hayward NK, Prins JB, Brown I, Walker N, et al. Leptin and the risk of Barrett's oesophagus. Gut. 2008;57:448–54.
111. Bhaskaran K, Douglas I, Forbes H, dos-Santos-Silva I, Leon DA, Smeeth L. Body-mass index and risk of 22 specific cancers: a population-based cohort study of 5·24 million UK adults. Lancet. 2014;384:755–65.
112. Steffen A, Huerta JM, Weiderpass E, Bueno-de-Mesquita HB, May AM, Siersema PD, et al. General and abdominal obesity and risk of esophageal and gastric adenocarcinoma in the European prospective investigation into cancer and nutrition. Int J Cancer. 2015;137:646–57.
113. Colquhoun A, Arnold M, Ferlay J, Goodman KJ, Forman D, Soerjomataram I. Global patterns of cardia and non-cardia gastric cancer incidence in 2012. Gut. 2015;64:1881–8.
114. Ravindran RD, Vashist P, Gupta SK, Young IS, Maraini G, Camparini M, et al. Prevalence and risk factors for vitamin C deficiency in north and south India: a two centre population based study in people aged 60 years and over. PLoS One. 2011;6:e28588.
115. Kong P, Cai Q, Geng Q, Wang J, Lan Y, Zhan Y, et al. Vitamin intake reduce the risk of gastric cancer: meta-analysis and systematic review of randomized and observational studies. PLoS One. 2014;9:e116060.
116. Katona P, Katona-Apte J. The interaction between nutrition and infection. Clin Infect Dis. 2008;46:1582–8.
117. Nishida T, Sakakibara H. Association between underweight and low lymphocyte count as an indicator of malnutrition in Japanese women. J Womens Health. 2010;19:1377–83.
118. Jo HH, Kim N, Jang J, Choi Y, Park JH, Park YM, et al. Impact of body mass index on survival depending on sex in 14,688 patients with gastric cancer in a tertiary hospital in South Korea. Gut Liver. 2023;17:243–58.
119. Ziomkiewicz A, Ellison PT, Lipson SF, Thune I, Jasienska G. Body fat, energy balance and estradiol levels: a study based on hormonal profiles from complete menstrual cycles. Hum Reprod. 2008;23:2555–63.
120. Key TJ, Appleby PN, Reeves GK, Roddam A, Dorgan JF, Longcope C, et al. Body mass index, serum sex hormones, and breast cancer risk in postmenopausal women. J Natl Cancer Inst. 2003;95:1218–26.
121. Lindblad M, Ye W, Rubio C, Lagergren J. Estrogen and risk of gastric cancer: a protective effect in a nationwide cohort study of patients with prostate cancer in Sweden. Cancer Epidemiol Biomarkers Prev. 2004;13:2203–7.
122. Sheh A, Ge Z, Parry NM, Muthupalani S, Rager JE, Raczynski AR, et al. 17beta-estradiol and tamoxifen prevent gastric cancer by modulating leukocyte recruitment and oncogenic pathways in *Helicobacter pylori*-infected INS-GAS male mice. Cancer Prev Res (Phila). 2011;4:1426–35.
123. Messa C, Pricci M, Linsalata M, Russo F, Di Leo A. Inhibitory effect of 17beta-estradiol on growth and the polyamine metabolism of a human gastric carcinoma cell line (HGC-27). Scand J Gastroenterol. 1999;34:79–84.
124. Ning S. Innate immune modulation in EBV infection. Herpesviridae. 2011;2:1.
125. Chen XZ, Chen H, Castro FA, Hu JK, Brenner H. Epstein-Barr virus infection and gastric cancer: a systematic review. Medicine (Baltimore). 2015;94:e792.
126. Burke AP, Yen TS, Shekitka KM, Sobin LH. Lymphoepithelial carcinoma of the stomach with Epstein-Barr virus demonstrated by polymerase chain reaction. Mod Pathol. 1990;3:377–80.
127. Bae JM, Kim EH. Epstein-Barr virus and gastric cancer risk: a meta-analysis with meta-regression of case-control studies. J Prev Med Public Health. 2016;49:97–107.
128. Camargo MC, Kim KM, Matsuo K, Torres J, Liao LM, Morgan DR, et al. Anti-*Helicobacter pylori* antibody profiles in 25 Epstein-Barr virus (EBV)- positive and EBV-negative gastric cancer. Helicobacter. 2016;21:153–7.

129. van Beek J, zur Hausen A, Klein Kranenbarg E, van de Velde CJ, Middeldorp JM, van den Brule AJ, et al. EBV-positive gastric adenocarcinomas: a distinct clinicopathologic entity with a low frequency of lymph node involvement. J Clin Oncol. 2004;22:664–70.
130. Li S, Du H, Wang Z, Zhou L, Zhao X, Zeng Y. Meta-analysis of the relationship between Epstein-1 Barr virus infection and clinicopathological features of patients with gastric carcinoma. Sci China Life Sci. 2010;53:524–30.
131. Li S, Murphy G, Pfeiffer R, Camargo MC, Rabkin CS. Meta-analysis shows that prevalence of Epstein-Barr virus-positive gastric cancer differs based on sex and anatomic location. Gastroenterology. 2009;137:824–33.
132. Kijima Y, Ishigami S, Hokita S, Koriyama C, Akiba S, Eizuru Y, et al. The comparison of the prognosis between Epstein–6 Barr virus (EBV)-positive gastric carcinomas and EBV-negative ones. Cancer Lett. 2003;200:33–40.
133. Camargo MC, Koriyama C, Matsuo K, Kim WH, Herrera-Goepfert R, Liao LM, et al. Case-case comparison of smoking and alcohol risk associations with Epstein-Barr virus-positive gastric cancer. Int J Cancer. 2014;134:948–53.
134. Kayamba V, Monze M, Asombang AW, Zyambo K, Kelly P. Serological response to Epstein-Barr virus early antigen is associated with gastric cancer and human immunodeficiency virus infection in Zambian adults: a case-control study. Pan Afr Med J. 2016;23:45.
135. Ignatova E, Seriak D, Fedyanin M, Tryakin A, Pokataev I, Menshikova S, et al. Epstein-Barr virus-associated gastric cancer: disease that requires special approach. Gastric Cancer. 2020;23:951–60.
136. Kim HJ, Hwang SW, Kim N, Yoon H, Shin CM, Park YS, et al. *Helicobacter pylori* and molecular markers as prognostic indicators for gastric cancer in Korea. J Cancer Preven. 2014;19:56–67.
137. Lee JH, Kim SH, Han SH, An JS, Lee ES, Kim YS. Clinicopathological and molecular characteristics of Epstein-Barr virus associated gastric carcinoma: a meta-analysis. J Gastroenterol Hepatol. 2009;24:354–65.
138. Tokunaga M, Land CE. Epstein-Barr virus involvement in gastric cancer: biomarker for lymph node metastasis. Cancer Epidemiol Biomarkers Prev. 1998;7:449–50.
139. Camargo MC, Kim WH, Chiaravalli AM, Kim KM, Corvalan AH, Matsuo K, et al. Improved survival of gastric cancer with tumour Epstein-Barr virus positivity: an international pooled analysis. Gut. 2014;63:236–43.
140. Tokunaga M, Uemura Y, Tokudome T, Ishidate T, Masuda H, Okazaki E, et al. Epstein-Barr virus related gastric cancer in Japan: a molecular patho-epidemiological study. Acta Pathol Jpn. 1993;43:574–81.
141. Sohn BH, Hwang JE, Jang HJ, Lee HS, Oh SC, Shim JJ, et al. Clinical significance of four molecular subtypes of gastric cancer identified by the cancer genome atlas project. Clin Cancer Res. 2017;23:4441–9.
142. Uemura Y, Tokunaga M, Arikawa J, Yamamoto N, Hamasaki Y, Tanaka S, et al. A unique morphology of Epstein-Barr virus related early gastric carcinoma. Cancer Epidemiol Biomarkers Prev. 1994;3:607–11.
143. Song HJ, Srivastava A, Lee J, Kim YS, Kim KM, Ki Kang W, et al. Host inflammatory response predicts survival of patients with Epstein-Barr virus-associated gastric carcinoma. Gastroenterology. 2010;139:84-92.e2.
144. Klein SL, Flanagan KL. Sex differences in immune responses. Nat Rev Immunol. 2016;16:626–38.
145. Araujo JM, Rosas G, Belmar-López C, Raez LE, Rolfo CD, Schwarz LJ, et al. Influence of sex in the molecular characteristics and outcomes of malignant tumors. Front Oncol. 2021;11:752918.
146. Clausen F, Behrens HM, Krüger S, Röcken C. Sexual dimorphism in gastric cancer: tumor-associated neutrophils predict patient outcome only for women. J Cancer Res Clin Oncol. 2020;146:53–66.
147. Haupt S, Caramia F, Klein SL, Rubin JB, Haupt Y. Sex disparities matter in cancer development and therapy. Nat Rev Cancer. 2021;21:393–407.

148. Klein SL, Jedlicka A, Pekosz A. The Xs and Y of immune responses to viral vaccines. Lancet Infect Dis. 2010;10:338–49.
149. Conforti F, Pala L, Pagan E, Bagnardi V, De Pas T, Queirolo P, et al. Sex-based dimorphism of anticancer immune response and molecular mechanisms of immune evasion. Clin Cancer Res. 2021;27:4311–24.
150. Polanczyk MJ, Hopke C, Vandenbark AA, Offner H. Estrogen-mediated immunomodulation involves reduced activation of effector T cells, potentiation of Treg cells, and enhanced expression of the PD-1 costimulatory pathway. J Neurosci Res. 2006;84:370–8.
151. Polanczyk MJ, Hopke C, Vandenbark AA, Offner H. Treg suppressive activity involves estrogen-dependent expression of programmed death-1 (PD-1). Int Immunol. 2007;19:337–43.
152. Guan X, Polesso F, Wang C, Sehrawat A, Hawkins RM, Murray SE, et al. Androgen receptor activity in T cells limits checkpoint blockade efficacy. Nature. 2022;606:791–6.
153. Kwon H, Schafer JM, Song NJ, Kaneko S, Li A, Xiao T, et al. Androgen conspires with the CD8^{+} T cell exhaustion program and contributes to sex bias in cancer. Sci Immunol. 2022;7:eabq2630.
154. Son HJ, Kim N, Song CH, Lee SM, Lee HN, Surh YJ. 17β-Estradiol reduces inflammation and modulates antioxidant enzymes in colonic epithelial cells. Korean J Intern Med. 2020;35:310–9.
155. Chakraborty B, Byemerwa J, Shepherd J, Haines CN, Baldi R, Gong W, et al. Inhibition of estrogen signaling in myeloid cells increases tumor immunity in melanoma. J Clin Invest. 2021;131:e151347.
156. Dong H, Zhu G, Tamada K, Chen L. B7–H1, a third member of the B7 family, co-stimulates T-cell proliferation and interleukin-10 secretion. Nat Med. 1999;5:1365–9.
157. Latchman Y, Wood CR, Chernova T, Chaudhary D, Borde M, Chernova I, et al. PD-L2 is a second ligand for PD-1 and inhibits T cell activation. Nat Immunol. 2001;2:261–8.
158. Carter L, Fouser LA, Jussif J, Fitz L, Deng B, Wood CR, et al. PD-1:PD-L inhibitory pathway affects both CD4(+) and CD8(+) T cells and is overcome by IL-2. Eur J Immunol. 2002;32:634–43.
159. Liang SC, Latchman YE, Buhlmann JE, Tomczak MF, Horwitz BH, Freeman GJ, et al. Regulation of PD-1, PD-L1, and PD-L2 expression during normal and autoimmune responses. Eur J Immunol. 2003;33:2706–16.
160. Mozaffarian N, Wiedeman AE, Stevens AM. Active systemic lupus erythematosus is associated with failure of antigen-presenting cells to express programmed death ligand-1. Rheumatology (Oxford). 2008;47:1335–41.
161. Nishimura H, Honjo T. PD-1: an inhibitory immunoreceptor involved in peripheral tolerance. Trends Immunol. 2001;22:265–8.
162. Nishimura H, Nose M, Hiai H, Minato N, Honjo T. Development of lupus-like autoimmune diseases by disruption of the PD-1 gene encoding an ITIM motif-carrying immunoreceptor. Immunity. 1999;11:141–51.
163. Han Y, Liu D, Li L. PD-1/PD-L1 pathway: current researches in cancer. Am J Cancer Res. 2020;10:727–42.
164. Correa P, Piazuelo MB. The gastric precancerous cascade. J Dig Dis. 2012;13:2–9.
165. Dai C, Geng R, Wang C, Wong A, Qing M, Hu J, et al. Concordance of immune checkpoints within tumor immune contexture and their prognostic significance in gastric cancer. Mol Oncol. 2016;10:1551–8.
166. McQuade JL, Daniel CR, Hess KR, Mak C, Wang DY, Rai RR, et al. Association of body-mass index and outcomes in patients with metastatic melanoma treated with targeted therapy, immunotherapy, or chemotherapy: a retrospective, multicohort analysis. Lancet Oncol. 2018;19:310–22.
167. Gu L, Chen M, Guo D, Zhu H, Zhang W, Pan J, et al. PD-L1 and gastric cancer prognosis: a systematic review and meta-analysis. PLoS One. 2017;12:e0182692.
168. Choi E, Chang MS, Byeon SJ, Jin H, Jung KC, Kim H, et al. Prognostic perspectives of PD-L1 combined with tumor-infiltrating lymphocytes, Epstein-Barr virus, and microsatellite instability in gastric carcinomas. Diagn Pathol. 2020;15:69.

169. Noh BJ, Kim JH, Eom DW. Prognostic significance of categorizing gastric carcinoma by PD-L1 expression and tumor infiltrating lymphocytes. Ann Clin Lab Sci. 2018;48:695–706.
170. Murphy G, Pfeiffer R, Camargo MC, Rabkin CS. Meta-analysis shows that prevalence of Epstein-Barr virus-positive gastric cancer differs based on sex and anatomic location. Gastroenterology. 2009;137:824–33.
171. Conforti F, Pala L, Bagnardi V, De Pas T, Martinetti M, Viale G, et al. Cancer immunotherapy efficacy and patients' sex: a systematic review and meta-analysis. Lancet Oncol. 2018;19:737–46.
172. Pala L, De Pas T, Catania C, Giaccone G, Mantovani A, Minucci S, et al. Sex and cancer immunotherapy: Current understanding and challenges. Cancer Cell. 2022;40:695–700.
173. Loo K, Tsai KK, Mahuron K, Liu J, Pauli ML, Sandoval PM, et al. Partially exhausted tumor-infiltrating lymphocytes predict response to combination immunotherapy. JCI Insight. 2017;2:e93433.
174. Layug PJ, Vats H, Kannan K, Arsenio J. Sex differences in $CD8^+$ T cell responses during adaptive immunity. WIREs Mech Dis. 2024;16:e1645.
175. Abramenko N, Vellieux F, Veselá K, Kejík Z, Hajduch J, Masařík M, et al. Investigation of the potential effects of estrogen receptor modulators on immune checkpoint molecules. Sci Rep. 2024;14:3043.
176. Castro A, Pyke RM, Zhang X, Thompson WK, Day CP, Alexandrov LB, et al. Strength of immune selection in tumors varies with sex and age. Nat Commun. 2020;11:4128.
177. Schreiber RD, Old LJ, Smyth MJ. Cancer immunoediting: integrating immunity's roles in cancer suppression and promotion. Science. 2011;331:1565–70.

Sexual Dimorphism in Hepatocellular Carcinoma Progression via ERα-Gα12 Signaling

Jihoon Tak and Sang Geon Kim

Abstract The heterotrimeric G protein alpha subunit 12 (Gα12) plays a role in regulating diverse cellular signaling pathways involved in physiological and pathological processes. Notably, Gα12 serves a critical modulator in the development and progression of hepatocellular carcinoma (HCC), the most common form of primary liver cancer with significant sex disparities in incidence and malignancy. Recent studies have demonstrated that Gα12 activation promotes oncogenic signaling that drives tumor cell proliferation, invasion, and metastasis, predominantly through induction of epithelial-mesenchymal transition (EMT). This transition facilitates cancer cell motility and dissemination, contributing to poor clinical outcomes. Intriguingly, Gα12 expression exhibits an inverse correlation with estrogen receptor alpha (ERα) levels, which are typically higher in female patients and confer a protective effect by suppressing Gα12 transcription. ERα-mediated signaling inhibits EMT markers and metastatic phenotypes, suggesting a mechanistic basis for sex-specific differences observed in HCC progression. Additionally, the interplay between Gα12 and ERα influences microRNA networks and downstream targets such as the phosphatase PTP4A1, further modulating tumor malignancy. Understanding the molecular crosstalk between Gα12 signaling and estrogen pathways provides valuable insights into the sexual dimorphism of liver cancer and may guide the development of targeted therapies aimed at disrupting Gα12-driven oncogenic processes while harnessing estrogen receptor-mediated tumor suppression.

J. Tak · S. G. Kim (✉)
College of Pharmacy, Dongguk University-Seoul, Goyang-si, Kyeonggi-do, Republic of Korea
e-mail: sgk@snu.ac.kr

J. Tak
Department of Anesthesiology, Critical Care and Pain Medicine, The University of Texas Health Science Center at Houston, McGovern Medical School, Houston, TX, USA

S. G. Kim
Department of Medicine, Graduate School, Dongguk University, Goyang-si, Kyeonggi-do, Republic of Korea

Biomedical Research Planning and Strategy for Dongguk University Medical Center, Goyang-si, Kyeonggi-do, Republic of Korea

H. Lee et al. (eds.), *Sex, Gender, and Emerging Technology in Healthcare: Mitigating Bias and Fostering Equity*, https://doi.org/10.1007/978-981-95-2070-1_3

Keywords Gα12 signaling pathway · Hepatocellular carcinoma · Estrogen receptor alpha · Epithelial-mesenchymal transition · Sex-specific cancer progression

1 Introduction

The objective of this chapter is to illustrate the occurrence and progression of sex-specific liver cancer associated with the estrogen receptor (ER)-dependent Gα12 signaling axis. Firstly, examining the onset and malignancy of liver cell cancer between sexes reveals noticeable sex-specific or sex-related differences [1, 2]. In South Korea, the incidence of liver cancer is approximately 3.5 times higher in males than in females, with a corresponding significant difference in mortality rates [3] (Fig. 1). Sex differences are reported in various liver disease models. For instance, in cases where the liver tissue is damaged by ischemic reperfusion injury, it has been reported that the damage is higher in females than in males, particularly observed in females under the age of 40 without macrosteatosis [4].

Currently, there is global attention on the progression of liver diseases, especially starting from simple hepatic steatosis, advancing through inflammation, liver cell damage, and progressing to fibrosis. When this fibrosis intensifies, leading to a more severe state, it results in cirrhosis [5]. Cirrhosis not only involves the pathological accumulation of fibrous tissue, but also signifies a situation where the actual

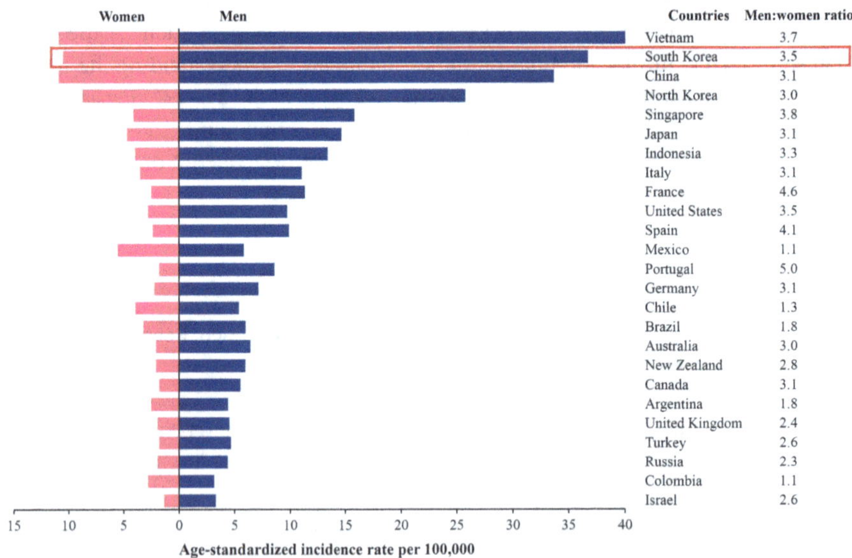

Fig. 1 Hepatocellular carcinoma (HCC) incidences showing sex disparity (adapted from Tang et al. [3])

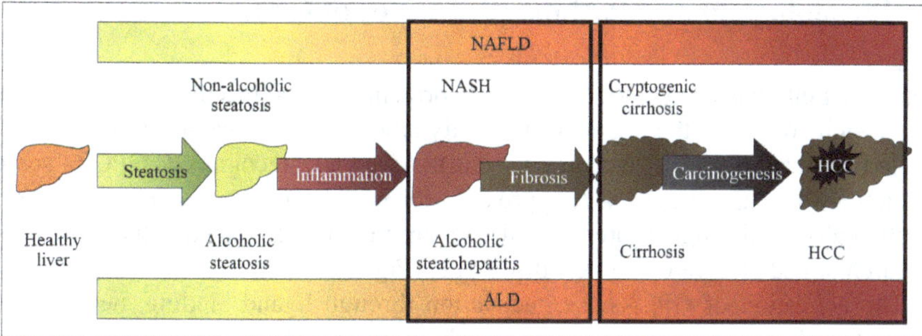

- ✓ Cluster of metabolic disorders
- ✓ ALD and NAFLD have very similar pathophysiological steps of disease progression
- ✓ Overweight, obesity, alcohol, physical inactivity, genetic factors and getting older
- ✓ A healthy lifestyle
  (not smoking, maintaining a healthy weight, limiting saturated fat, salt, alcohol)

Fig. 2 A scheme showing progression of metabolic liver syndrome and alcoholic liver disease (ALD) to HCC (adapted from Mahli et al. [7])

functioning liver cells decrease, leading to impaired liver function [6]. In patients with liver cirrhosis, there is an increased risk of developing hepatocellular carcinoma (HCC), emphasizing the need for early diagnosis and intervention to treat the disease [7] (Fig. 2).

The reason for paying attention to non-alcoholic steatohepatitis (NASH) and fibrosis in simple hepatic steatosis among various liver diseases is because non-alcoholic fatty liver disease (NAFLD), also called metabolic dysfunction-associated steatotic liver disease (MASLD), is increasingly recognized as the major cause, progressing from mild to severe conditions starting from hepatic steatosis [8]. Another cause of liver disease is alcoholic liver disease (ALD), resulting from chronic excessive alcohol consumption. While there are molecular differences between these two conditions, common pathological features such as dysregulation of autophagy as well as simple fat accumulation in hepatocytes are also involved [9].

Thus, the intensification of liver diseases, accompanied by inflammation, is gaining recent attention as a cause of metabolic disorders. Factors contributing to metabolic syndrome, such as overweight, excessive alcohol consumption, decreased physical activity, and genetic factors, when progressing with aging, can lead to the exacerbation of liver diseases and the occurrence of inflammatory hepatitis (NASH) [5, 7].

2 Physiological and Pathological Roles of Gα12 Axis

Sex-specificity has been associated with G protein-coupled receptors (GPCRs) [10, 11], which constitute the largest gene family among cell surface receptors, and the related signal transducer (e.g., Gα protein). As widely recognized, GPCRs serve as targets for many drugs, with approximately 40% of direct and indirect targets of clinically used drugs. Moreover, GPCRs are the most frequently studied area in research aimed at developing new drugs [12] (Fig. 3).

The activation of GPCRs, i.e., activation through ligand binding, necessitates a process of transmitting activation signals into the cell. In this process, GPCRs utilize the G protein complex, which is a component of the heterotrimeric complex. Additionally, GPCRs act as cell surface receptors, and this applies to multicellular organisms like mammals, including humans. GPCRs function as sensors that perceive changes in the external or internal environments, and the activation of signals downstream from GPCRs typically involves stimulation from different cell surface receptors [13].

In mammals, there are approximately 800 different types of GPCRs that may act as sensors for recognizing changes in environmental and internal stimuli. However, when these signals converge and enter the cell, a mediator or transducer is required to facilitate this process. Gα proteins have been studied as transducers since they play a crucial role in transmitting signals from these cell surface receptors into the cell, often considered a molecular bottleneck. Through this mediation, signals are further propagated through amplification, leading to various activation processes within the cell, or dampening for feedback regulation [13, 14].

In our research laboratory, particular attention has been given to the Gα12 family members, Gα12, and Gα13, among the GPCR-associated proteins. The major branches of G alpha proteins include Gαs, Gαi, Gαq, and Gα12 [15].

The GPCR-associated Gα12 signaling axis, as summarized in the above diagram, involves representative signal transduction pathways such as RhoA, ROCK1/2, and

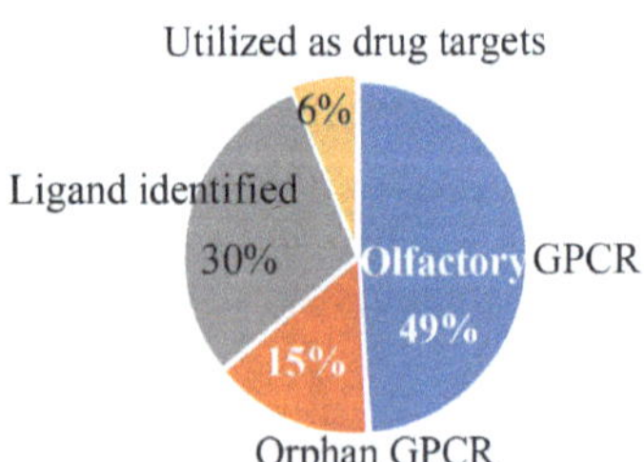

Fig. 3 G protein-coupled receptor (GPCR) and their physiological and pharmacological functions

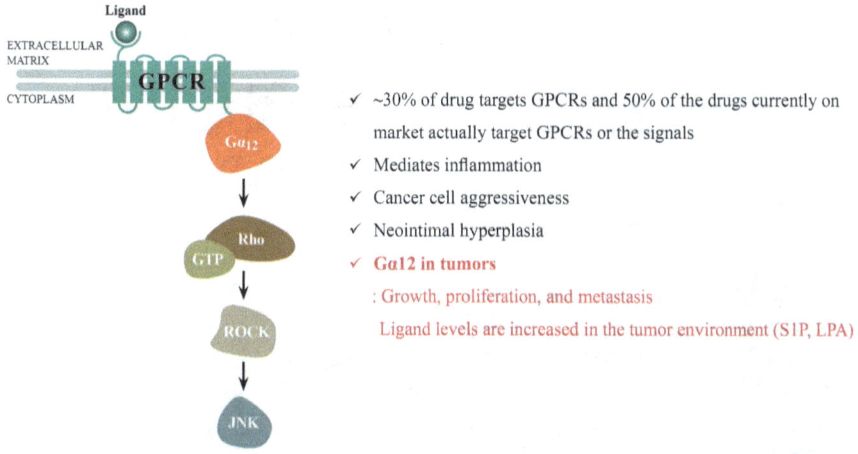

Fig. 4 Activated GPCR coupling to Gα12 and its representative downstream signal axis (modified from Yang et al. [16] and Ki et al. [17])

JNK (Fig. 4). These processes are associated with physiological responses or pathological conditions in the body. Examples include inflammatory responses, malignancy of cancer cells, and intimal hyperplasia, which are relevant to biological reactions in various situations. Particularly in cancer research, evidences suggest the implication of Gα12 in the proliferation and metastasis of different kinds of cancers [16–19].

The research focused on GPCRs that couple with Gα12 to activate this signaling axis reveals various ligands. These ligands include substances associated with the cancer microenvironment and liver diseases, or fibrosis and cirrhosis. Examples of such ligands are angiotensin II, lysophosphatidic acid (LPA), thrombin, endothelin A, and sphingosine-1-phosphate (S1P) [20–24]. Similar to the ligands of GPCR, the angiotensin II type 1 receptor, LPA-1 receptor, protease-activated receptor 1, endothelin A receptor, and S1P receptor are also known to be involved in the progression of liver fibrosis [25–29]. These molecules are considered stimuli that not only explain directly or indirectly the initiation and malignancy of cancer but also contribute to severe conditions such as fibrotic liver damage and related diseases (Fig. 5).

In our research laboratory, we have been investigating the physiological and pathological phenomena influenced by changes in Gα12 levels for two decades. In an early study from Dr. M Simmons' laboratory, examining the expression levels of Gα12/Gα13 in various tissues reveals common expression across multiple organs in the body, with notably high expression in the liver and skeletal muscle [30]. This suggests the potential involvement of these proteins in metabolic regulation processes in these organs.

As an example of our research, we have explored the relevance of Gα12/Gα13 in the context of fuel consumption. In animal studies, we observed changes in the

Liver fibrosis & GPCR-Gα12 signaling

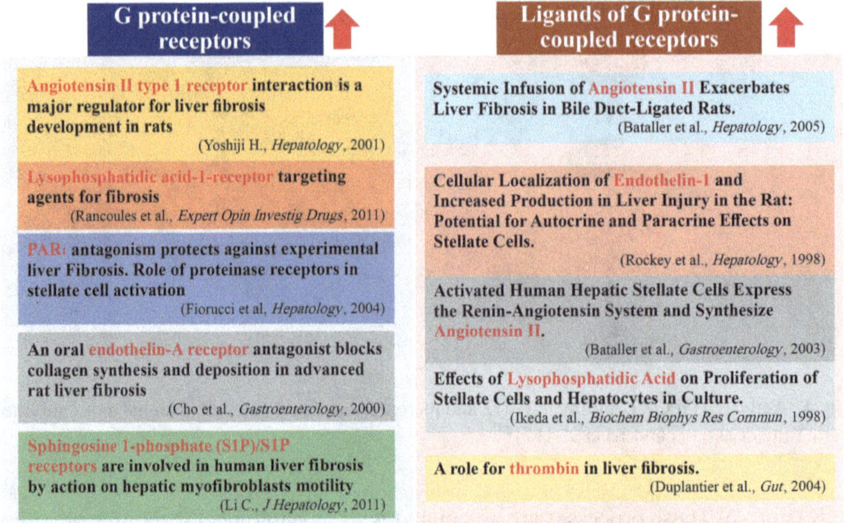

Fig. 5 Physiological responses associated with liver fibrosis activated by representative ligands and their GPCR-Gα12 signaling

expression of Gα12/Gα13 in skeletal muscles when comparing sedentary conditions to endurance exercise training. An important finding of our study is that Gα13 expression was decreased in endurance exercise training in association with myofiber type switch to oxidative forms. Morphological changes observed using electron microscopy also revealed a significant increase in the quantity of intermyofibrillar and subsarcolemmal mitochondria in Gα13 gene knockout animals [31]. This suggests a role for Gα13 in mitochondrial dynamics in muscle tissue. Simultaneously, there was an increase in Gα12 expression, indicating the involvement of Gα12 in fuel and oxygen consumption after a shift to oxidative myofiber types.

Apart from studies on muscle tissue changes, we have also studied the physiological role of Gα12 axis in liver tissue. The model used in our liver study was the fasting-refeeding model, where animals subjected to fasting exhibited an increase in Gα12 expression after 24–48 h. Specifically, free fatty acids mobilized from adipose tissue move through the bloodstream and accumulate in the liver during fasting, undergoing conversion to neutral fats, resulting in fat accumulation. Interestingly, the expression of Gα12 protein and its mRNA increased during the 24–48 h following fasting. To explore the role of Gα12 in this context, a genetic model was employed. Notably, when Gα12 was genetically removed, there was a further increase in the accumulation of neutral fats in the liver tissue induced by fasting. This was evident both morphologically, by observing the amount of fat accumulating in the liver during fasting, and biochemically, through increased intensity in Oil Red O staining, confirming the findings through biochemical analysis or histological examination [32].

Having identified the impact of Gα12 signaling axis on fatty acid oxidation, we tried to find the target genes regulated by the Gα12 signaling axis. When comparing the results between the cells with genetically activated Gα12 and the wild-type cell line, core proteins such as PPARα, PGC1α, and SIRT1 were found to be centrally involved in the network of fatty acid oxidation genes [32]. Particularly, this network involves lipid catabolism, ketogenesis, and peroxisomal oxidation protein groups along with mitochondrial biogenesis, supporting the role of Gα12 in mitochondrial fatty acids oxidation.

Specifically, when animals with a genetic deficiency in Gα12 were subjected to a long-term high-fat diet, there was a higher accumulation of neutral fats compared to normal animals on a high-fat diet. This was accompanied by increased body weight gain and an increase in fat mass and size [32]. This phenomenon is interpreted as a manifestation of impaired metabolic activation.

In this case, interest arises in identifying the GPCR(s) associated with the Gα12 signaling axis, so we have investigated ADP and adenosine as potential ligands for GPCRs coupled to Gα12/Gα13 and examined the possibility of these ligands serving as indicators of energy depletion as they are ATP metabolites. ADP stimulates the GPCR P2Y13, and adenosine binds to four types of GPCR: A1, A2a, A2b, and A3. No significant role for Gα12 was found in the P2Y13 study. However, there is a significant increase in adenosine levels during fasting, presumably originating from the liver [32], and the coupling of adenosine receptors to Gα12.

In cell line studies, it has been elucidated that Gα12 protein is involved in the regulation of SIRT1 mediated by USP22 under adenosine receptor subtypes. When combining these results, it becomes clear that Gα12 plays a central role in the binding of adenosine to the receptors and signal transduction processes, with the recognition of adenosine as an indicator of energy waste activating Gα12-coupled GPCRs.

In addition to its physiological functions, the overexpression of Gα12 seems to be involved in the stress response of liver tissue when exposed to external harmful factors such as overdose of drugs and toxic substances. Endoplasmic reticulum (ER) stress, as one form of external stress, is part of the pathological process involving Gα12. For instance, in cases of damage caused by excessive intake of liver-intoxicating drugs like acetaminophen, ER stress occurs. This triggers an overexpression of Gα12 through a process mediated by IRE1α. In this context, Gα12 is transcriptionally activated by Xbp1s downstream from IRE1α, and thereby the protein level is increased through gene activation [33].

Overexpressed Gα12, in turn, inhibits miR-15a in hepatocytes through a pathway mediated by ROCK1. This leads to an increase in ALOX12, resulting in hepatocyte necrosis, where the death pathway involved is ferroptosis [33]. The entire process indicates that Gα12 plays a role in the liver tissue's response to stress induced by external factors, particularly ER stress in the context of drug-induced liver injury (Fig. 6).

The process where hepatocyte cell death leads to inflammation and, concurrently, activation of hepatic stellate cells (HSCs) resulting in fibrosis is a well-known phenomenon. HSCs serve as the primary storage site for retinol in the body, maintaining it in the form of retinyl esters within lipid droplets. Upon liver injury, these

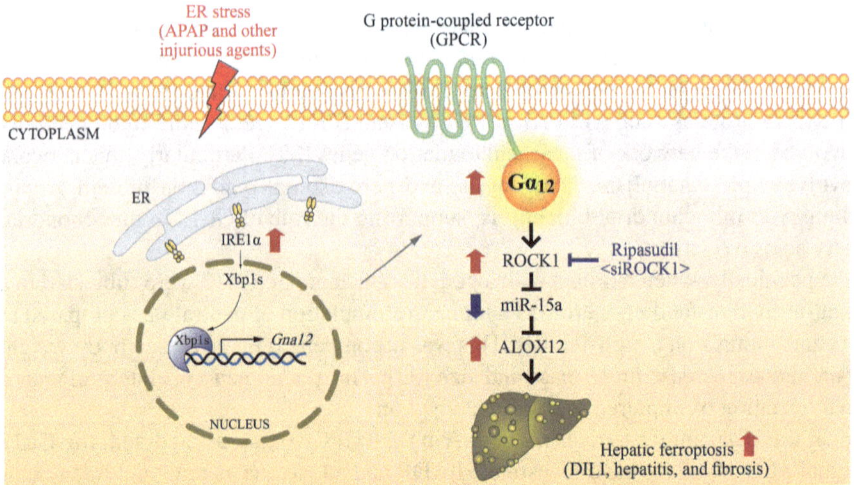

Fig. 6 Gα12 overexpression and endoplasmic reticulum (ER) stress in hepatocytes exacerbates acute liver injury via ROCK1-mediated miR-15a and ALOX12 dysregulation (adapted from Tak et al. [33])

cells release large amounts of stored retinol, a portion of which is converted into retinoic acid [34]. This retinoid-mediated modulation of autophagic activity could have significant health implications, as autophagy plays a role in recycling damaged or unnecessary cellular components. Beyond the well-established Gα12-activating ligands, such as angiotensin II, S1P, and thrombin [35–37], autophagy can also be stimulated by PDGF and TGFβ [38, 39]. Within this framework of autophagy induction and Gα12 signaling, recent studies have demonstrated that Gα12 overexpression in HSCs enhances autophagy, thereby accelerating the progression of liver fibrosis [40] (Fig. 7).

In our research, we have observed a significant decrease in miR-16 in the process, leading to the de-repression of the de novo synthesis process of Gα12 that was previously inhibited by miR-16. In quiescent HSCs (i.e., normal, non-activated state), miR-16 levels are high. In this process, miR-16 inhibits the synthesis by binding to the 3′-UTR of Gα12 mRNA [40], and this inhibition is released when miR-16 is lowered under ER stress conditions. Hence, upon activation due to stress, the decrease in miR-16 level becomes involved in the protein synthesis process of Gα12.

Moreover, Gα12 increase contributes to the activation of autophagy during this period, promoting the proliferation and activation of HSCs [40]. This process is linked to ER stress, showing a dual role where ER stress is involved in hepatocytes cell death and in activation and proliferation of HSCs simultaneously. Ultimately, the increased activation of GPCR signaling, which recognizes external signals, amplifies autophagy (Fig. 8). In summary, these research findings contribute to the understanding of the physiological and pathological roles of Gα12, with implications not only in liver-related diseases but also in cancer research.

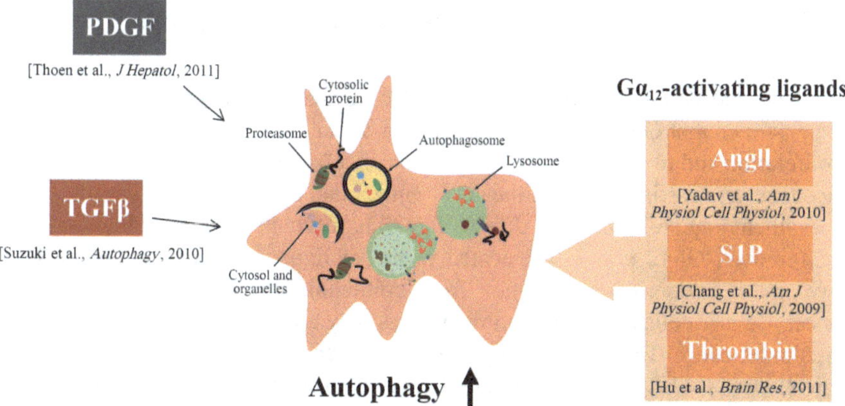

Fig. 7 Overexpression of Gα12 promotes autophagy in hepatic stellate cells (HSCs) in response to cell surface ligands facilitating fibrosis

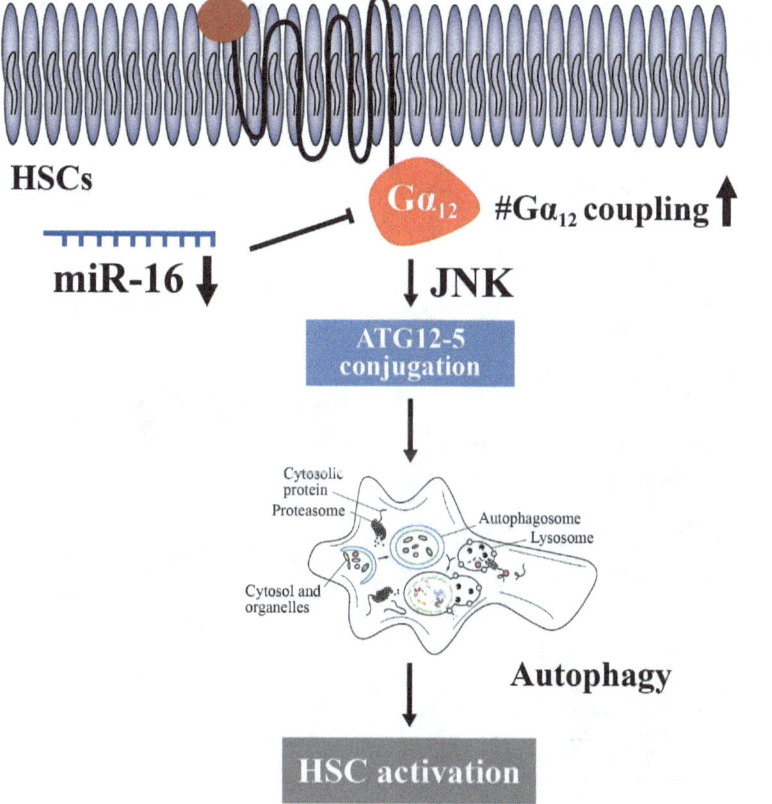

Fig. 8 Dysregulation of miR-16 elicits Gα12 overexpression, which activates HSCs by facilitating autophagy via ATG12-5 conjugation (modified from Kim et al. [40])

3 Roles of Gα12 in HCC Malignancy and Sex Specificity

Another important finding from our research on Gα12, as depicted in Fig. 9, we have demonstrated that Gα12 is involved in various signal transductions that promote the proliferation and metastasis of cancer. In cancer cells, Gα12 is designated as the *gep* oncogene, and it participates in the development and proliferation of various cancer types, including HCC, through diverse signaling mechanisms [16, 41].

Moreover, Gα12 is also implicated in the malignancy of cancer. When HCC develops, it undergoes a transition from epithelial cancer to mesenchymal-type cancer. Cancer cells that undergo this process play a key role in the movement and invasion of cancer from the primary lesion. In other words, enhanced expression or activation of Gα12, whether through overexpression or the formation of an active mutant, augments this pathway, thereby contributing to enhanced cancer progression and invasion [16, 41, 42] (Fig. 9).

The heatmap in the figure published in the International Journal of Cancer illustrates a significant increase in Gα12 expression when HCC becomes malignant. In the case of metastatic cancer, the expression of *GNA12* (Gα12) is notably higher compared to non-metastatic cancer tissues, not to mention adjacent normal tissues (Fig. 10). More importantly, the highest inverse correlation existed between ESR1 (ERα) and Gα12 expression. Interestingly, when ERα expression is high, Gα12

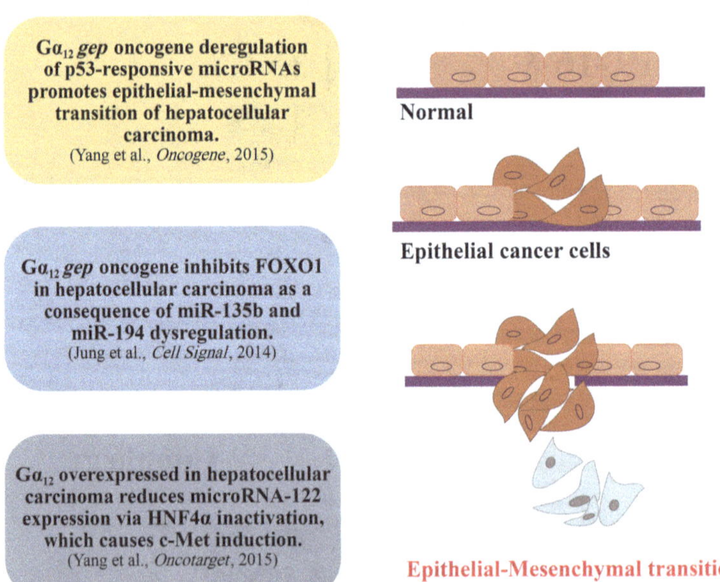

Fig. 9 Gα12 protein works as a *gep* oncogene in liver cancer, promoting epithelial-to-mesenchymal transition (EMT)

expression tends to be low. This correlation was stronger compared to the correlation with ESR2 (ERβ) or G protein-coupled estrogen receptor (GPER) [43].

What adds further intrigue is that patients with low ERα expression and high Gα12 expression in the cancer tissues exhibited lower overall survival rates and recurrence-free survival rates. These changes in markers within patient tissues are supported by immunoblotting or tissue immunohistochemistry (IHC) analysis [43].

The receptors to which 17β-estradiol binds are distinctly identified as ERα, ERβ, and GPER, as depicted in the Fig. 11. According to various reports, when several regulatory factors such as miR-18 and Erbin [44, 45] inhibit ERα, the activation of various cancers, including HCC, is promoted [46]. Interestingly, single nucleotide polymorphisms (SNPs) at FOXA2 binding sites have been shown to impair the

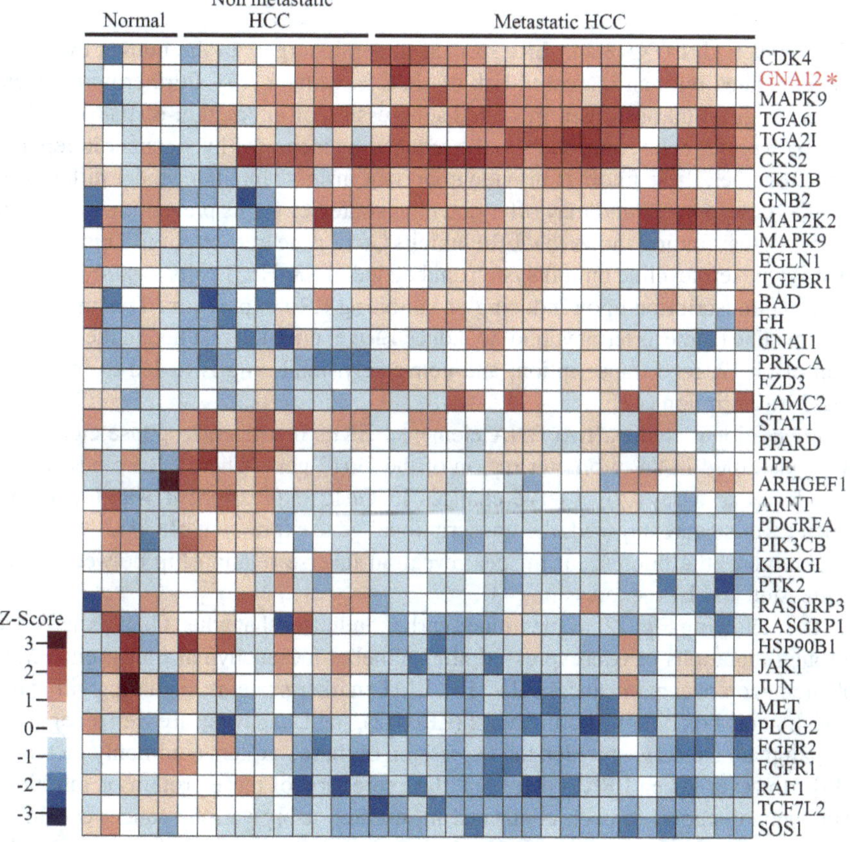

Fig. 10 Heat map showing significantly changed mRNAs across the progression of HCC using the GEO database (adapted from Yun et al. [43])

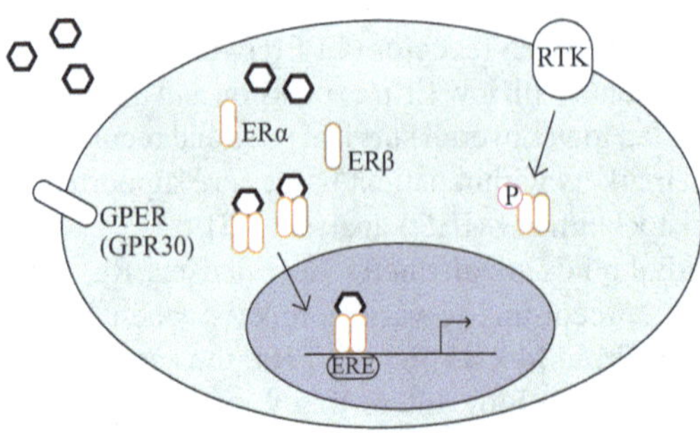

Focused on sex hormones, especially ERα

- miR-18 → ERα inhibition
  Liu WH et al., *Gastroenterology*, 2009
- Erbin → ERα inhibition
  Hua WH et al., *Journal of hepatology*, 2017
- ERα with Foxa1/2 → HCC inhibition
  Li Z et al., *Cell*, 2012
- IL-6 secretion inhibition
  Willscott E. N et al., *Science*, 2007

Fig. 11 A scheme showing ERα binding to ERE in the target genes and representative reports on the outcomes of physiological modulations of ERα

binding of both FOXA2 and ERα to their genomic targets in the human liver, and are associated with an increased risk of HCC in women [47]. Furthermore, recent studies suggest that estrogen-mediated suppression of interleukin-6 (IL-6) production by Kupffer cells contributes to the reduced incidence of liver cancer in females [48]. Together, these findings provide important insights into the gender differences in HCC development and may offer potential strategies for its prevention in humans (Fig. 11). Our finding shows that ERα inhibits Gα12 gene transcription. This research utilized the tetracycline on–off system embedded in SK-Hep1 cells to study changes in Gα12 transcription upon inducing or inhibiting ERα expression. Treatment with doxycycline (same as tetracycline) to induce ERα expression resulted in the suppression of Gα12 gene transcription, and conversely, inhibiting ERα led to increased Gα12 transcription [43].

In additional studies, two DNA elements, EREs (estrogen response elements), were identified in the promoter region of the Gα12 gene where ERα binds. Using reporter gene assays to observe changes in Gα12 expression, it was found that when ERα is overexpressed, the transcriptional gene expression of Gα12 decreases. Furthermore, the fluorescent images reveal an inverse relationship between Gα12 expression and ERα expression [43].

The following Fig. 12, as mentioned earlier, indicates that when Gα12 is activated through mutations or overexpressed, the epithelial-mesenchymal transition (EMT) phenomenon occurs in cancer cells [16]. EMT marker proteins, including vimentin, zinc finger E-box binding homeobox 1 (ZEB1/2), snail, and twist, are well recognized in malignant cancers, in which the expression of these markers commonly increases [49]. However, when ERα is activated, the expression of these proteins is inhibited. This phenomenon has been confirmed in research using various cancer cell lines, including SNU449, HepG2, SK-Hep1, and SNU387 cells [43]. In summary, the activation of ERα appears to suppress the expression of EMT marker proteins in cancer cells, as demonstrated across different cell lines.

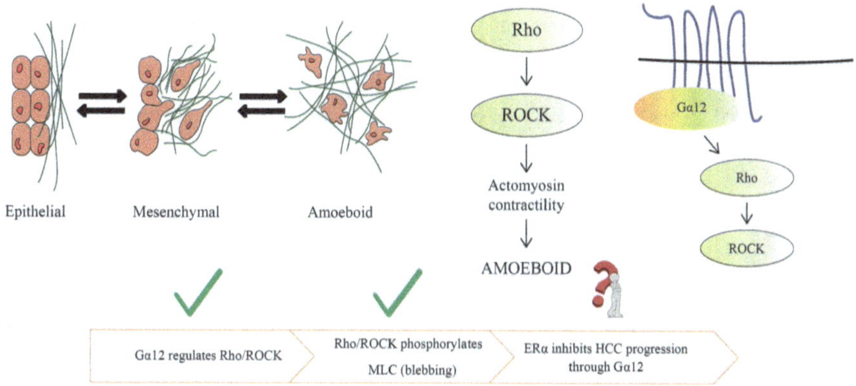

Fig. 12 Role of Gα12 as a facilitator of amoeboid cancer cell movement in HCC (modified from Yun et al. [43])

In our previous study, we conducted cDNA microarray analysis to identify tumor suppressor microRNAs (miRNAs) following overexpression of an active mutant of Gα12 (Gα12QL). Through further research analyzing the miRNAs we discovered, it was determined that miR-141 and miR-200a are regulated by ERα. In the process, when cancer cell lines were treated with estrogen, miR-141 and miR-200a increased, as mentioned earlier. Conversely, inhibiting ERα using siRNA decreased the expression of miR-141 and miR-200a [43].

Furthermore, when SNU449 and HepG2 cell lines were directly treated with 100 nM concentration of estrogen, the level of miR-141 and miR-200a were observed to increase over time. Conversely, the opposite phenomenon was observed when ERα was inhibited with siRNA [43].

It can be inferred that various targets of downstream signaling regulated by ERα and Gα12 may play functional roles. Among them, what we focused on are phosphatases, enzymes that have been studied as targets in cancer, which could be a new target regulated by Gα12, prompting in-depth research. Since there were no specific reports on PTP4A1 among phosphatases (Fig. 13), we conducted additional research on its correlation with Gα12. Through preliminary studies, we found that PTP4A1 interacts with the miRNA we derived, and further research reveals the binding of PTP4A1 3'-UTR with miR-141, as well as miR-200-3p binding to the same seed sequence. Functionally, as predicted, treating with miR-141 and miR-200a mimics results in decreased PTP4A1 protein expression, while inhibition leads to the opposite effect. Similarly, treating cancer cells with estrogen or expressing ERα leads to a decrease in PTP4A1.

Previously, we mentioned the importance of EMT in cancer malignancies. Overexpression of PTP4A1 is associated with a decrease in E-cadherin along with an increase in vimentin, supporting the occurrence of EMT. Conversely, inhibiting PTP4A1 results in the reversal of these phenomena, supporting the correlation between PTP4A1 and EMT [43].

ERα/Gα12 downstream target

1. Phosphatases importance

✓ PTPs are emerging new targets for drug discovery

(*Cell*, 2017 and *BBA*, 2013)

✓ PTP4A1~3 are validated oncology targets

(*Cancer Res*, 2008)

2. Known to function in cell proliferation, metastasis, and tumor angiogenesis

(*Cancer Res*, 2003)

Fig. 13 PTP4A1 as a new downstream target from the ERα/Gα12 signaling axis

In addition to EMT, another consideration in the metastatic process of cancer cells is that primary tumor cells at the primary site need to infiltrate and move out through the mesh of connective tissue abundantly accumulated around cancer cells, requiring penetration and migration into the bloodstream. Furthermore, when extravasation from the bloodstream to other secondary sites for metastasis, similar changes may be necessary [49]. This process involves the morphologically flexible change in cancer cells through the phenomenon known as mesenchymal-amoeboid transition (MAT), which is of course associated with EMT [50]. In this context, we conducted additional research on the roles of ERα and Gα12.

In morphological studies and immunocytochemistry analysis, it has been observed that activating ERα or treating with estrogen significantly reduces ameboid movement. Conversely, when Gα12 is removed, the opposite reversal of this phenomenon has been confirmed. Moreover, activation of Gα12 leads to an increase in p-MLC, a marker of the MAT, and overexpression of ERα or high levels of estrogen result in the inhibition of MAT, as indicated by a decrease in p-MLC. This supports the hypothesis that overexpression of ERα and high estrogen levels can hinder the malignancy of cancer through the inhibition of MAT. Additionally, it has been confirmed that the activation of ERα during this process inhibits the GTP-RhoA signaling pathway, which is a downstream signal of the Gα12 pathway (Figs. 12 and 13) [43].

Furthermore, we confirmed whether these observations are evident in human samples through correlation analysis in the assays of liver cancer tissues. We also analyzed whether this phenomenon ultimately manifests itself in specific HCC malignancy. When analyzing liver cancer samples supplied by Asan Medical Center (Seoul, South Korea), we observed a common elevation of PTP4A1 and p-MLC in cancer lesions compared to surrounding tissues. Additionally, further analysis of these samples for sex differences showed that Gα12, PTP4A1, and p-MLC were all lower in women compared to men. In addition, miR-141 and miR-200a were consistently higher in women compared to men in these subgroups [43]. Based on the findings from our laboratory and other information [51], the evidence for females having resistance to the malignancy of HCC is interpreted to originate from the relatively higher levels of estrogen female hormones (Fig. 14). We believe that this sex-specific phenomenon will contribute to explaining the substantial differences in

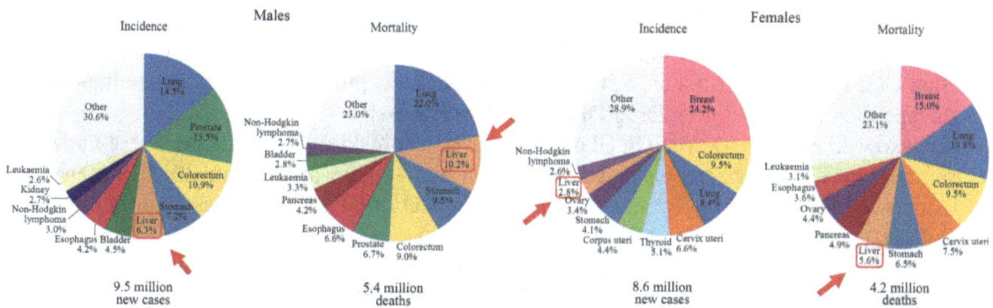

Fig. 14 The incidence and mortality of male and female patients with HCC (adapted from Bray et al. [51])

the HCC incidence between men and women, as well as differences in the degree of malignancy after the onset of liver cancer.

Acknowledgements SGK's external stipend was supported in part by the National Research Foundation (NRF) grant funded by the Korea government (MSIT) (RS-2024-00441114 and RS-2024-00454443)

References

1. Zhu C, Boutros PC. Sex differences in cancer genomes: much learned, more unknown. Endocrinology. 2021;162.
2. Liu P, Xie SH, Hu S, Cheng X, Gao T, Zhang C, et al. Age-specific sex difference in the incidence of hepatocellular carcinoma in the United States. Oncotarget. 2017;8:68131–7.
3. Tang A, Hallouch O, Chernyak V, Kamaya A, Sirlin CB. Epidemiology of hepatocellular carcinoma: target population for surveillance and diagnosis. Abdom Radiol (NY). 2018;43:13–25.
4. Han S, Cho J, Wi W, Won Lee K, Hwa Cha H, Lee S, et al. Sex difference in the tolerance of hepatic ischemia-reperfusion injury and hepatic estrogen receptor expression according to age and macrosteatosis in healthy living liver donors. Transplantation. 2022;106:337–47.
5. Lazarus JV, Mark HE, Anstee QM, Arab JP, Batterham RL, Castera L, et al. Advancing the global public health agenda for NAFLD: a consensus statement. Nat Rev Gastroenterol Hepatol. 2022;19:60–78.
6. Berasain C, Arechederra M, Argemí J, Fernández-Barrena MG, Avila MA. Loss of liver function in chronic liver disease: an identity crisis. J Hepatol. 2023;78:401–14.
7. Mahli A, Hellerbrand C. Alcohol and obesity: a dangerous association for fatty liver disease. Dig Dis. 2016;34(Suppl 1):32–9.
8. Chan WK, Chuah KH, Rajaram RB, Lim LL, Ratnasingam J, Vethakkan SR. Metabolic dysfunction-associated steatotic liver disease (MASLD): a state-of-the-art review. J Obes Metab Syndr. 2023;32:197–213.
9. Osna NA, Donohue TM Jr, Kharbanda KK. Alcoholic liver disease: pathogenesis and current management. Alcohol Res. 2017;38:147–61.
10. Broughton BR, Brait VH, Kim HA, Lee S, Chu HX, Gardiner-Mann CV, et al. Sex-dependent effects of G protein-coupled estrogen receptor activity on outcome after ischemic stroke. Stroke. 2014;45:835–41.

11. Mouat MA, Coleman JLJ, Smith NJ. GPCRs in context: sexual dimorphism in the cardiovascular system. Br J Pharmacol. 2018;175:4047–59.
12. Blad CC, Tang C, Offermanns S. G protein-coupled receptors for energy metabolites as new therapeutic targets. Nat Rev Drug Discov. 2012;11:603–19.
13. Trzaskowski B, Latek D, Yuan S, Ghoshdastider U, Debinski A, Filipek S. Action of molecular switches in GPCRs–theoretical and experimental studies. Curr Med Chem. 2012;19:1090–109.
14. Congreve M, de Graaf C, Swain NA, Tate CG. Impact of GPCR structures on drug discovery. Cell. 2020;181:81–91.
15. Worzfeld T, Wettschureck N, Offermanns S. G(12)/G(13)-mediated signalling in mammalian physiology and disease. Trends Pharmacol Sci. 2008;29:582–9.
16. Yang YM, Lee WH, Lee CG, An J, Kim ES, Kim SH, et al. Gα12 gep oncogene deregulation of p53-responsive microRNAs promotes epithelial-mesenchymal transition of hepatocellular carcinoma. Oncogene. 2015;34:2910–21.
17. Ki SH, Choi MJ, Lee CH, Kim SG. Galpha12 specifically regulates COX-2 induction by sphingosine 1-phosphate. Role for JNK-dependent ubiquitination and degradation of IkappaBalpha. J Biol Chem. 2007;282:1938–47.
18. Kang KW, Choi SY, Cho MK, Lee CH, Kim SG. Thrombin induces nitric-oxide synthase via Galpha12/13-coupled protein kinase C-dependent I-kappaBalpha phosphorylation and JNK-mediated I-kappaBalpha degradation. J Biol Chem. 2003;278:17368–78.
19. Kim YM, Lim SC, Han CY, Kay HY, Cho IJ, Ki SH, et al. G(alpha)12/13 induction of CYR61 in association with arteriosclerotic intimal hyperplasia: effect of sphingosine-1-phosphate. Arterioscler Thromb Vasc Biol. 2011;31:861–9.
20. Bataller R, Gäbele E, Parsons CJ, Morris T, Yang L, Schoonhoven R, et al. Systemic infusion of angiotensin II exacerbates liver fibrosis in bile duct-ligated rats. Hepatology. 2005;41:1046–55.
21. Rockey DC, Fouassier L, Chung JJ, Carayon A, Vallee P, Rey C, et al. Cellular localization of endothelin-1 and increased production in liver injury in the rat: potential for autocrine and paracrine effects on stellate cells. Hepatology. 1998;27:472–80.
22. Bataller R, Sancho-Bru P, Ginès P, Lora JM, Al-Garawi A, Solé M, et al. Activated human hepatic stellate cells express the renin-angiotensin system and synthesize angiotensin II. Gastroenterology. 2003;125:117–25.
23. Ikeda H, Yatomi Y, Yanase M, Satoh H, Nishihara A, Kawabata M, et al. Effects of lysophosphatidic acid on proliferation of stellate cells and hepatocytes in culture. Biochem Biophys Res Commun. 1998;248:436–40.
24. Duplantier JG, Dubuisson L, Senant N, Freyburger G, Laurendeau I, Herbert JM, et al. A role for thrombin in liver fibrosis. Gut. 2004;53:1682–7.
25. Yoshiji H, Kuriyama S, Yoshii J, Ikenaka Y, Noguchi R, Nakatani T, et al. Angiotensin-II type 1 receptor interaction is a major regulator for liver fibrosis development in rats. Hepatology. 2001;34:745–50.
26. Rancoule C, Pradère JP, Gonzalez J, Klein J, Valet P, Bascands JL, et al. Lysophosphatidic acid-1-receptor targeting agents for fibrosis. Expert Opin Investig Drugs. 2011;20:657–67.
27. Fiorucci S, Antonelli E, Distrutti E, Severino B, Fiorentina R, Baldoni M, et al. PAR1 antagonism protects against experimental liver fibrosis. Role of proteinase receptors in stellate cell activation. Hepatology. 2004;39:365–75.
28. Cho JJ, Hocher B, Herbst H, Jia JD, Ruehl M, Hahn EG, et al. An oral endothelin—a receptor antagonist blocks collagen synthesis and deposition in advanced rat liver fibrosis. Gastroenterology. 2000;118:1169–78.
29. Li C, Zheng S, You H, Liu X, Lin M, Yang L, et al. Sphingosine 1-phosphate (S1P)/S1P receptors are involved in human liver fibrosis by action on hepatic myofibroblasts motility. J Hepatol. 2011;54:1205–13.
30. Strathmann MP, Simon MI. G alpha 12 and G alpha 13 subunits define a fourth class of G protein alpha subunits. Proc Natl Acad Sci U S A. 1991;88:5582–6.
31. Koo JH, Kim TH, Park SY, Joo MS, Han CY, Choi CS, et al. Gα13 ablation reprograms myofibers to oxidative phenotype and enhances whole-body metabolism. J Clin Invest. 2017;127:3845–60.

32. Kim TH, Yang YM, Han CY, Koo JH, Oh H, Kim SS, et al. Gα12 ablation exacerbates liver steatosis and obesity by suppressing USP22/SIRT1-regulated mitochondrial respiration. J Clin Invest. 2018;128:5587–602.
33. Tak J, Kim YS, Kim TH, Park GC, Hwang S, Kim SG. Gα(12) overexpression in hepatocytes by ER stress exacerbates acute liver injury via ROCK1-mediated miR-15a and ALOX12 dysregulation. Theranostics. 2022;12:1570–88.
34. Haaker MW, Vaandrager AB, Helms JB. Retinoids in health and disease: a role for hepatic stellate cells in affecting retinoid levels. Biochim Biophys Acta Mol Cell Biol Lipids. 2020;1865: 158674.
35. Yadav A, Vallabu S, Arora S, Tandon P, Slahan D, Teichberg S, et al. ANG II promotes autophagy in podocytes. Am J Physiol Cell Physiol. 2010;299:C488–96.
36. Chang CL, Ho MC, Lee PH, Hsu CY, Huang WP, Lee H. S1P(5) is required for sphingosine 1-phosphate-induced autophagy in human prostate cancer PC-3 cells. Am J Physiol Cell Physiol. 2009;297:C451–8.
37. Hu S, Xi G, Jin H, He Y, Keep RF, Hua Y. Thrombin-induced autophagy: a potential role in intracerebral hemorrhage. Brain Res. 2011;1424:60–6.
38. Thoen LF, Guimarães EL, Dollé L, Mannaerts I, Najimi M, Sokal E, et al. A role for autophagy during hepatic stellate cell activation. J Hepatol. 2011;55:1353–60.
39. Suzuki HI, Kiyono K, Miyazono K. Regulation of autophagy by transforming growth factor-β (TGF-β) signaling. Autophagy. 2010;6:645–7.
40. Kim KM, Han CY, Kim JY, Cho SS, Kim YS, Koo JH, et al. Gα(12) overexpression induced by miR-16 dysregulation contributes to liver fibrosis by promoting autophagy in hepatic stellate cells. J Hepatol. 2018;68:493–504.
41. Yang YM, Lee CG, Koo JH, Kim TH, Lee JM, An J, et al. Gα12 overexpressed in hepatocellular carcinoma reduces microRNA-122 expression via HNF4α inactivation, which causes c-Met induction. Oncotarget. 2015;6:19055–69.
42. Jung HS, Seo YR, Yang YM, Koo JH, An J, Lee SJ, et al. Gα12gep oncogene inhibits FOXO1 in hepatocellular carcinoma as a consequence of miR-135b and miR-194 dysregulation. Cell Signal. 2014;26:1456–65.
43. Yun J, Kim YS, Heo MJ, Kim MJ, Moon A, Kim SG. ERα inhibits mesenchymal and amoeboidal movement of liver cancer cell via Gα12. Int J Cancer. 2022;150:1690–705.
44. Liu WH, Yeh SH, Lu CC, Yu SL, Chen HY, Lin CY, et al. MicroRNA-18a prevents estrogen receptor-alpha expression, promoting proliferation of hepatocellular carcinoma cells. Gastroenterology. 2009;136:683–93.
45. Wu H, Yao S, Zhang S, Wang JR, Guo PD, Li XM, et al. Elevated expression of Erbin destabilizes ERα protein and promotes tumorigenesis in hepatocellular carcinoma. J Hepatol. 2017;66:1193–204.
46. Prossnitz ER, Barton M. Estrogen biology: new insights into GPER function and clinical opportunities. Mol Cell Endocrinol. 2014;389:71–83.
47. Li Z, Tuteja G, Schug J, Kaestner KH. Foxa1 and Foxa2 are essential for sexual dimorphism in liver cancer. Cell. 2012;148:72–83.
48. Naugler WE, Sakurai T, Kim S, Maeda S, Kim K, Elsharkawy AM, et al. Gender disparity in liver cancer due to sex differences in MyD88-dependent IL-6 production. Science. 2007;317:121–4.
49. Kalluri R, Weinberg RA. The basics of epithelial-mesenchymal transition. J Clin Invest. 2009;119:1420–8.
50. Ren L, Mo W, Wang L, Wang X. Matrine suppresses breast cancer metastasis by targeting ITGB1 and inhibiting epithelial-to-mesenchymal transition. Exp Ther Med. 2020;19:367–74.
51. Bray F, Ferlay J, Soerjomataram I, Siegel RL, Torre LA, Jemal A. Global cancer statistics 2018: GLOBOCAN estimates of incidence and mortality worldwide for 36 cancers in 185 countries. CA Cancer J Clin. 2018;68:394–424.

Part III
Sex Differences in Neuroscience and Brain Health

Sex-Specific Insights in Neuroscience: From Brain Structure to Mental Health Interventions

Heajin Kim and Heisook Lee

Abstract Sex differences in neuroscience hold critical implications for understanding mental health disparities and tailoring interventions. This review examines biological and sociocultural factors influencing sex-specific variations in brain structure, function, and neurophysiology, with a focus on disorders such as Alzheimer's, depression, autism spectrum disorder, and pain. Advances in neuroimaging, genomics, and big data analytics reveal distinct patterns in brain morphology, connectivity, and neurotransmitter activity, offering insights into sex-based susceptibilities to mental disorders. Microglia, the brain's resident immune cells, exhibit sex-dependent roles in neuroinflammation and synaptic plasticity, underscoring their relevance in neurological diseases. Despite progress, challenges remain, including underrepresentation in clinical studies and the lack of standardized methodologies. This review advocates for integrating sex as a biological variable in neuroscience research to enhance the quality of research. Interdisciplinary collaboration and sex-inclusive research designs are essential for addressing these disparities and improving global health outcomes.

Keywords Neuroscience · Sex differences · Microglia · Neurodevelopmental disorders · Neurodegenerative diseases · Mental health · Precision medicine

H. Kim (✉)
Policy Research Team, Korea Center for Gendered Innovations in Science and Technology Research, Seoul, Korea
e-mail: khj826@gister.re.kr; khj826@hanmail.net

H. Lee
President, Korea Center for Gendered Innovations in Science and Technology Research, Seoul, Korea

Professor Emeritus, Ewha Womans University, Seoul, Korea

H. Lee et al. (eds.), *Sex, Gender, and Emerging Technology in Healthcare: Mitigating Bias and Fostering Equity*, https://doi.org/10.1007/978-981-95-2070-1_4

1 Understanding Sex Differences in Neuroscience: Challenges and Opportunities

Sex differences in neuroscience have emerged as a crucial area of research, highlighting distinct biological and sociocultural factors that influence brain structure, function, and susceptibility to mental health disorders. Historically, much neuroscience research treated males and females as interchangeable, often overlooking sex as a critical biological variable. This oversight limited our understanding of neurological and psychiatric conditions with clear sex-specific patterns, such as the higher prevalence of dementia in women and the greater incidence of autism spectrum disorder (ASD) in men. Addressing these disparities requires a nuanced exploration of the mechanisms underlying sex differences, supported by rigorous methodological approaches.

In recent years, neuroscience has undergone a paradigm shift, increasingly recognizing sex as a critical biological variable rather than a confounding factor. A growing body of evidence demonstrates that biological sex significantly influences neuroanatomical development, neurotransmitter dynamics, and vulnerability to neurological and psychiatric disorders. This evolving recognition has prompted methodological reforms, such as the adoption of the "Sex as a Biological Variable (SABV)" framework, and policy mandates by funding agencies and scientific journals to promote sex-inclusive study designs. Technological innovations—including advanced neuroimaging, machine learning, and high-throughput genomics—have enabled the detection of nuanced sex-based patterns in the brain that were previously obscured. Despite these advances, challenges persist: sex-disaggregated data are still inconsistently reported, and analytical rigor in interpreting sex differences remains uneven. Nevertheless, the field is moving toward a more holistic, integrative understanding of brain function that accounts for both biological and sociocultural dimensions of sex and gender. This trend reflects a broader commitment to precision medicine and equitable health outcomes across sexes.

One significant barrier to progress has been the historical exclusion of female subjects in preclinical studies. This practice was largely justified by the assumption that hormonal fluctuations in females increased variability in experimental results. However, a meta-analysis [1] challenged this belief by analyzing over 6,000 data points across behavioral, electrophysiological, neurochemical, and histological measures. The study found no significant sex differences in variability, thus refuting the assumption of instability and advocating for balanced inclusion of both sexes in neuroscience research. Despite these findings, a decade-long review of neuroscience and psychiatry papers [2, 3] revealed that only 19% of studies employed optimal designs to analyze sex differences, and a mere 5% treated sex as a discovery variable. This highlights the need for methodological advancements to reveal nuanced, sex-specific insights.

Technological progress has begun to address these gaps. Advances in neuroimaging, genomics, and big data analytics have enabled researchers to map sex differences in brain morphology, connectivity, and neurotransmitter dynamics.

For example, studies on brain structure have identified differences in the corpus callosum, anterior cingulate cortex, and hippocampus, regions critical for cognitive and emotional processing [4, 5]. Functional studies further reveal distinct patterns of brain activation [5, 6], with male brains exhibiting stronger intra-hemispheric connectivity supporting motor and sensory functions, while female brains show greater inter-hemispheric connectivity associated with integrative and analytical processes. These findings underscore that sex differences extend beyond size variations, reflecting intrinsic biological distinctions shaped by hormonal and genetic factors.

Emerging evidence from behavioral neuroscience also emphasizes the importance of sex-inclusive approaches. The SABV initiative has highlighted biases in male-centric experimental paradigms. For instance, classical behavioral assays often fail to account for hormonal cycles in females, producing results that are either not replicable or less informative for understanding sex-specific mechanisms. The present study critiques these traditional paradigms and underscores the value of integrating SABV in experimental designs to capture behavioral diversity across sexes [7]. By adopting SABV, researchers have identified new mechanisms underlying neurological and psychiatric disorders, such as sex-specific microglial responses to neuroinflammation in conditions like Alzheimer's disease [8].

Despite these advancements, challenges remain. Methodological biases, such as the underrepresentation of females in research and inadequate analytical frameworks, continue to hinder progress. However, emerging technologies, particularly machine learning and advanced neuroimaging techniques, offer promising solutions. These tools can refine behavioral analyses, improve data interpretation, and uncover nuanced patterns of sex-specific differences, paving the way for more equitable and precise neuroscience research [9].

This paper examines the biological basis of sex differences in the brain, focusing on their implications for mental health. By reviewing current findings on microglial functions, neurotransmitter systems, and structural and functional disparities, we aim to elucidate the mechanisms driving these differences. Furthermore, we advocate for the systematic integration of SABV in neuroscience research, advancing precision medicine and equitable healthcare solutions.

2 Biological Basis of Sex Differences

2.1 Structural Differences

Recent advances in neuroimaging, computational modeling, and big data analytics have illuminated significant structural differences between male and female brains [5, 10–15]. These distinctions are rooted in both biological and sociocultural factors, offering critical insights into sex-specific neural organization and cognitive adaptations. While historical research often conflated biological sex and socially constructed gender, contemporary studies emphasize the need to disentangle these influences to accurately interpret brain morphology.

Using high-resolution imaging techniques, researchers have consistently documented distinct patterns in brain size, shape, and connectivity. For example, a deep learning-based study analyzed over 1,000 diffusion MRI scans and achieved 93.3% accuracy in distinguishing male and female brains [5] (Fig. 1). The analysis revealed that male brains exhibited greater intra-hemispheric connectivity, which enhances motor coordination and sensory processing, while female brains displayed stronger inter-hemispheric connectivity, facilitating integrative and analytical functions. These findings align with earlier diffusion tensor imaging studies, which observed similar connectivity patterns in adolescents and young adults [12], suggesting that these differences become more pronounced during adolescence due to divergent neural maturation processes.

Gray matter distribution also varies independently of brain size. The study demonstrated that women exhibit greater gray matter volumes in the prefrontal cortex and other regions associated with emotional regulation and verbal processing [16]. These differences persist even when brain size is matched between sexes, indicating intrinsic biological dimorphism rather than simple size effects.

The study, which separated the effects of biological sex and gender traits on brain morphology [13], found that female sex correlated with larger subcortical volumes, including those of the basal ganglia, hippocampus and ventral diencephalon. In contrast, gender traits, measured along a continuum, were associated with cortical thickness in prefrontal regions such as the superior frontal and caudal middle frontal cortices. Individuals scoring higher on feminine traits exhibited greater cortical thickness in these areas, regardless of biological sex. This distinction underscores the importance of considering sex and gender as semi-independent variables to avoid conflating biological and sociocultural influences.

Large-scale studies contextualize these structural differences within evolutionary and social frameworks. The "10,000 Social Brains" study, which utilized UK Biobank data, highlighted sex-specific adaptations in the limbic system, a brain region critical for emotional regulation and social interactions [4]. Women's limbic volumes, such as the amygdala and ventromedial prefrontal cortex, correlated strongly with enriched social environments, suggesting neural specialization for nurturing and empathetic roles. In contrast, men's brain structures reflected adaptations to hierarchical and competitive dynamics. These findings align with the social brain hypothesis, which posits that neural differences evolved to support distinct sex-specific social behaviors.

Machine learning techniques, including convolutional neural networks, have enhanced the precision of sex classification and revealed complex interregional patterns previously inaccessible through traditional methods [5]. Key discriminative regions include the corpus callosum, anterior corona radiata, and anterior cingulate cortex—areas implicated in cognitive and emotional regulation. These tools not only improve the accuracy of identifying neural dimorphism but also reveal complex patterns that traditional methods might overlook.

The structural differences between male and female brains have profound implications for understanding sex-specific susceptibilities to neurological and psychiatric

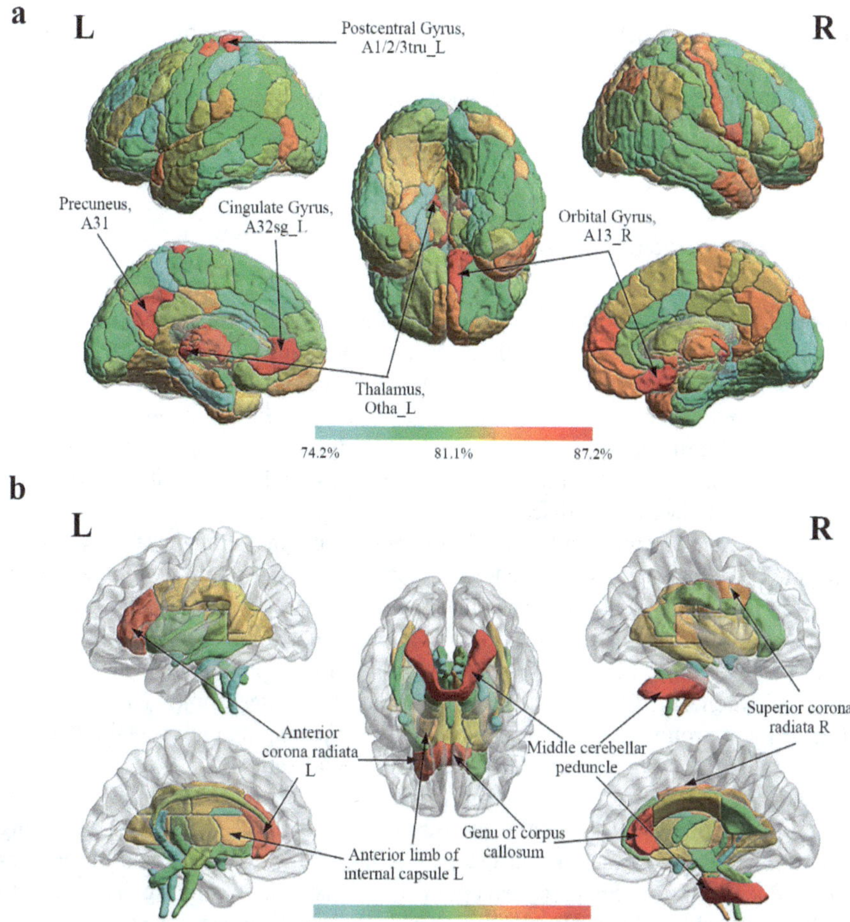

Fig. 1 Classification accuracy in brain gray and white matter regions (adapted from Xin et al. [5]). Brain imaging analysis identified the regions most effective in distinguishing between groups based on classification accuracy. **a** Gray matter: using the human brainnetome atlas, results from 246 Gy matter regions showed the highest classification accuracy in the following areas: left precuneus (BA 31)—87.2%, Left postcentral gyrus (BA 1/2/3, trunk region)—87.2%, left subgenual cingulate gyrus (BA 32)—87.2%, Right orbital gyrus of the frontal lobe (BA 13)—87.1%, Left occipital thalamus—86.9%, **b** white matter: according to the ICBM-DTI-81 white-matter labels atlas, results from 48 white matter regions showed top accuracy in: middle cerebellar peduncle—89.7%, Genu of the corpus callosum—88.4%, Right anterior corona radiate—88.3%, Right superior corona radiate—86.0%, Left anterior limb of the internal capsule—85.4%

disorders [3, 6, 17–19]. For instance, women's larger gray matter volumes in emotion-related regions may explain their higher prevalence of depression and anxiety, while men's enhanced intra-hemispheric connectivity aligns with their strengths in spatial tasks and higher rates of ASD.

2.2 Functional Differences

Sex-specific differences in brain function manifest through distinct patterns of neural activity, connectivity, and adaptability. These functional disparities are rooted in biological and environmental factors, influencing cognitive processes, behavioral outcomes, and susceptibilities to neurological and psychiatric conditions. Recent research combining neuroimaging, behavioral analysis, and computational methods has advanced our understanding of these differences, highlighting their complexity and relevance to health and disease [20, 21].

Studies reveal that male and female brains employ distinct strategies for processing information, supported by differences in connectivity. The study which analyzed diffusion MRI data from 949 individuals aged 8–22, showed that male brains are optimized for intra-hemispheric communication, enhancing motor coordination and sensory processing [12]. In contrast, female brains emphasize inter-hemispheric communication, facilitating analytical and intuitive integration. These patterns become more pronounced during adolescence, suggesting that hormonal influences and neural maturation contribute to functional specialization.

Early life studies further underscore functional differences. Female neonates exhibit more widespread cortical responses to painful stimuli than males, reflecting early-onset sex differences in sensory processing pathways [22]. Such differences may influence long-term pain perception and processing, offering insights into sex-based variability in conditions like chronic pain and anxiety disorders.

The capacity for neural plasticity, or the brain's ability to reorganize itself structurally and functionally, also shows notable sex-specific patterns. In conditions such as ASD, where the male-to-female diagnosis ratio is approximately 4:1, these differences become particularly significant. Males are more susceptible to disruptions in genes governing synaptic plasticity, which are often linked to ASD [23–25]. These disruptions affect highly variable brain regions, such as the associative cortex, which is integral to perception and social cognition.

Research further demonstrates sex-specific plasticity responses to injury or disease. For example, functional cortical plasticity following brain lesions, such as traumatic brain injury or multiple sclerosis, differs by sex [26–28]. Males are more prone to synaptic disruptions, while females exhibit greater resilience, possibly due to hormonal influences that enhance adaptability in certain brain regions [29–31]. These findings underscore the importance of considering SABV when developing therapeutic interventions for neurological disorders.

2.3 Neurotransmitter Dynamics

Sex-specific differences in neurotransmitter systems are critical to understanding how male and female brains regulate cognitive, emotional, and physiological processes. These differences influence susceptibility to neurological and psychiatric disorders, including pain, addiction, and neurodevelopmental conditions. Recent advances in neurochemical analysis, pharmacological studies, and neuroimaging have illuminated how neurotransmitter systems operate differently in males and females, revealing distinct molecular and functional pathways.

Pain regulation is one of the most well-documented areas of sex-specific neurotransmitter dynamics. Neurotransmitters such as glutamate, γ-aminobutyric acid (GABA), serotonin, dopamine, and noradrenaline interact with key brain regions like the periaqueductal gray (PAG), amygdala, and nucleus accumbens (Nac) to mediate pain perception and response.

A pivotal study found that female mice exhibit higher expression of prostaglandin D-synthase, leading to enhanced prostaglandin production and altered pain responses compared to males [32]. This molecular divergence underpins differences in pain sensitivity, with females generally experiencing higher pain sensitivity and lower thresholds for certain types of pain.

Additionally, research on opioid analgesia has shown sex-specific responses to drugs like morphine. Female rats demonstrate lower sensitivity to morphine and develop analgesic tolerance more rapidly than males. This finding is linked to differences in neurotransmitter receptor activity in pain-regulating regions such as the PAG and Nac. The study revealed that chronic opioid exposure causes location-specific and sex-dependent changes in receptor signaling, with males exhibiting greater facilitation at thalamostriatal synapses [33]. These insights emphasize the necessity for sex-specific pain management strategies.

Neurotransmitter dynamics also play a crucial role in reward systems, where sex differences have profound implications for addiction research and treatment. Dopamine, a key regulator of reward and motivation, operates differently in male and female brains. Males exhibit reduced hyperlocomotion in response to amphetamines, while females show calming effects [34]. These behavioral differences correlate with variations in dopamine transporter expression, highlighting distinct dopaminergic regulatory mechanisms.

Further research into opioid addiction underscores these sex-specific differences. Male mice show increased GABA release probability and impaired inhibitory plasticity in the ventral tegmental area (VTA) during withdrawal, contributing to heightened withdrawal symptoms [35]. Conversely, female mice maintain baseline GABA activity and exhibit greater resilience to synaptic impairments. These findings suggest that sex-tailored therapies could improve treatment outcomes for opioid use disorders.

Serotonin, a neurotransmitter critical for mood regulation, exhibits notable sex differences [36, 37]. Women have higher serotonin levels in the blood and greater expression of 5-HT1A(serotonin 1A) receptors in cortical and subcortical regions compared to men. However, the rate of serotonin synthesis is approximately 52% faster in men. These differences may partly explain sex-based disparities in mood disorders such as depression and anxiety, which are more prevalent in women.

Dopamine, which regulates reward and motor systems, is also influenced by sex hormones [19, 34, 35, 38, 39]. Female hormones like β-estradiol enhance dopamine cell activity and release, contributing to sex-specific patterns in motor coordination, addiction, and neurodegenerative conditions such as Parkinson's disease [19]. For instance, women's greater dopamine release in the presence of estrogen may offer protective effects against Parkinson's disease, though this advantage diminishes post-menopause.

GABA, the brain's primary inhibitory neurotransmitter, also shows significant sex differences, particularly during development [35, 40–42]. Estrogen modulates GABA receptor efficacy, resulting in differences in synaptic activity between males and females [35, 42]. This dynamic is especially relevant in neurodevelopmental disorders such as ASD, where GABAergic disruptions are implicated.

Sex differences in neurotransmitter systems have profound implications for understanding and treating neurological and psychiatric disorders [19, 25, 26, 32, 34–38, 43–46] (Table 1).

To fully leverage these findings, future research should integrate sex as a variable in studies of neurotransmitter systems. Longitudinal analyses can track changes in neurotransmitter activity across the lifespan, particularly in response to hormonal

Table 1 Sex differences in neurotransmitter systems and disorder type

Disorder type	Sex-specific characteristics	Sex-specific considerations
Pain disorders [32, 43, 44]	Females exhibit higher expression of prostaglandin D-synthase, leading to greater pain sensitivity and lower thresholds	Females show reduced morphine sensitivity and develop tolerance more rapidly; pain management strategies must be sex-specific
Addiction [34, 35]	Males show increased GABA release and impaired inhibitory plasticity in the VTA during withdrawal; females show greater synaptic resilience	Sex-tailored therapies should consider heightened withdrawal symptoms in males and enhanced synaptic recovery in females
Mood disorders [36–39, 45]	Women have higher serotonin levels and 5-HT1A receptor expression; serotonin synthesis is faster in men	Treatment should address women's higher susceptibility to depression and anxiety, incorporating hormonal fluctuations
Neurodevelopmental disorders [19, 25, 26, 46]	Males are more susceptible to GABAergic disruptions during development, influencing ASD vulnerability	Therapeutic strategies should account for estrogen's modulation of GABA and dopamine systems in females

GABA γ-aminobutyric acid, *5-HT1A* serotonin 1A, *ASD* autism spectrum disorder, *VTA* ventral tegmental area fluctuations or environmental factors. Advanced imaging techniques, combined with computational models, can further elucidate the interplay between neurotransmitters and sex hormones in regulating brain function. These efforts will not only enhance our understanding of sex-specific brain dynamics but also pave the way for precision medicine approaches tailored to the unique neurochemical profiles of men and women.

2.4 *Sex-Specific Microglial Functions*

Microglia, the central nervous system's resident immune cells, play a critical role in maintaining neural health through diverse functions, including immune surveillance, synaptic remodeling, and response to injury [26, 47]. These dynamic cells are highly adaptable, responding to both intrinsic factors, such as genetic and hormonal influences, and extrinsic factors, including environmental changes and disease states [8, 48, 49]. Recent research underscores that microglia are not homogeneous across sexes; rather, they exhibit distinct behaviors and functional capacities in males and females, influenced by hormonal regulation, genetic variations, and life-stage transitions [9, 27, 50, 51] (Fig. 2).

During embryogenesis and early postnatal life, microglia are central to brain development, ensuring proper neural circuitry formation by pruning excess synapses and regulating neural progenitor cell proliferation. While sex differences in microglial function are subtle during early gestation, they become more pronounced in late fetal stages [8, 49, 52]. For instance, studies in mice reveal that female microglia exhibit heightened sensitivity to inflammatory stimuli, such as lipopolysaccharides, at late gestation (E18.4). In contrast, male microglia show greater motility and responsiveness shortly after birth, reflecting divergent developmental trajectories.

Sex differences in microglial function during early brain development are strongly linked to the risk of neurodevelopmental disorders. ASD is 4–5 times more prevalent in males [23, 53, 54], and this disparity is partially attributed to male microglia's inflammatory bias. During critical windows of neural development, excessive synaptic pruning and heightened immune activation by male microglia can disrupt neural circuitry formation, contributing to ASD pathology. In contrast, female microglia exhibit reparative and regulatory tendencies, mitigating risks associated with dysregulated pruning and inflammation [50] Similar to ASD, attention deficit hyperactivity disorder (ADHD) is more common in males, potentially due to microglial-driven alterations in neural connectivity. Male microglia's heightened immune responses during early development can impair the formation of attention-related neural networks.

In adulthood, microglia continue to perform essential maintenance functions, interacting with astrocytes and oligodendrocyte progenitor cells to support neural health and repair. However, sex differences remain evident in microglial activity and molecular profiles [50, 51, 55]. Male microglia exhibit a pronounced pro-inflammatory bias, characterized by elevated expression of immune-related proteins

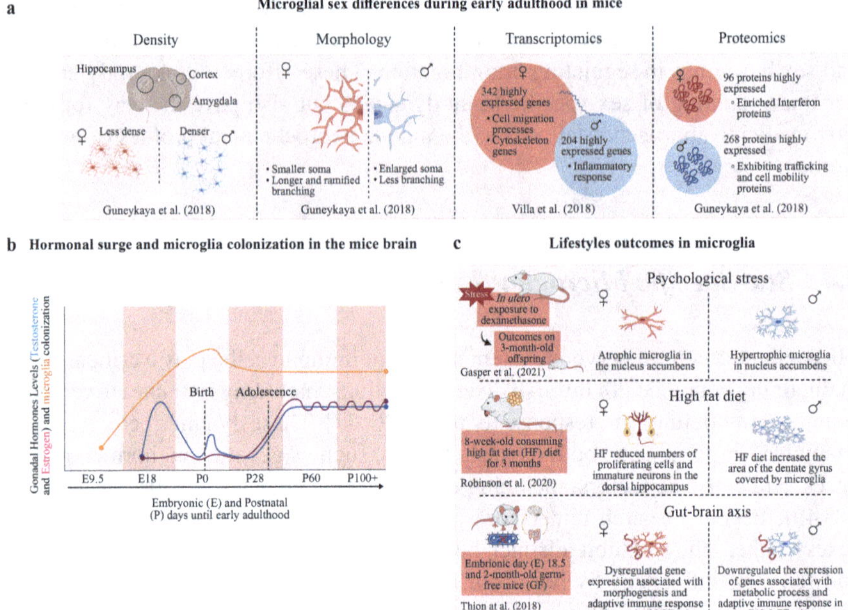

Fig. 2 Key differences in microglial properties between sexes (adapted from Bobotis et al. [27]) **a** Microglia, the brain's resident immune cells, show distinct differences between males and females. In females, microglia typically have smaller cell bodies and more branched processes. Molecular studies also reveal that gene and protein expression patterns vary by sex. Additionally, microglial density differs, with males having higher levels in certain brain regions. **b** These differences begin early in development. In mice, microglia start colonizing the brain as early as embryonic day 9.5, coinciding with rising gonadal hormone levels. Both hormone activity and microglial populations stabilize around postnatal day 30 and remain consistent into adulthood. **c** Environmental and lifestyle factors also influence microglia in sex-specific ways. Psychological stress causes microglial shrinkage (atrophy) in females and enlargement (hypertrophy) in males. A high-fat diet increases microglial density in males but reduces cell proliferation and accelerates neuron loss in females. Furthermore, the absence of gut microbiota disrupts different gene pathways in each sex—affecting immune responses and brain structure in females, and metabolism in males

such as major histocompatibility complex I (MHCI) and MHCII. This predisposition to inflammation increases susceptibility to neuroinflammatory damage and may contribute to the progression of conditions such as Parkinson's disease and traumatic brain injury. By contrast, female microglia demonstrate enhanced phagocytic activity and neuroprotective capabilities, underpinned by higher expression of genes associated with repair and regeneration [8].

Hormonal regulation further accentuates these differences. Estrogen and progesterone modulate microglial function in females, enhancing anti-inflammatory responses and synaptic repair. These protective effects are particularly pronounced during reproductive years, reducing the risk of conditions like multiple sclerosis in women compared to men during this stage of life [9, 28, 49].

Sex-specific microglial behaviors are critical determinants of susceptibility and progression in neurodegenerative disorders, which exhibit pronounced sex differences in prevalence and severity. Women account for approximately two-thirds of Alzheimer's Disease cases, with faster disease progression and greater tau pathology compared to men [8, 9, 17, 19]. Female microglia transition to pro-inflammatory states earlier, particularly post-menopause, when protective effects of estrogen decline. This shift exacerbates tau accumulation and amyloid deposition, key hallmarks of Alzheimer's Disease. The apolipoprotein E4 (*APOE4*) allele, a major genetic risk factor for Alzheimer's Disease, disproportionately impacts female microglia, amplifying their inflammatory and neurodegenerative responses [8, 17, 30, 56].

Parkinson's Disease is more prevalent and severe in males, largely due to the inflammatory predisposition of male microglia [6, 8, 28, 57, 58]. Chronic inflammation driven by male microglia accelerates dopaminergic neuron loss, the hallmark of Parkinson's disease. Female microglia's neuroprotective functions may delay disease onset, though these effects diminish with aging as hormonal influences wane.

Multiple Sclerosis has a higher incidence in women, but disease severity is often greater in men. Female microglia initially exhibit robust reparative mechanisms, including enhanced phagocytosis and myelin repair. However, under chronic inflammation, female microglia can become dysregulated, contributing to disease progression. Male microglia exacerbate neural damage through persistent inflammatory signaling, hindering recovery and repair.

The distinct trajectories of microglial behavior in males and females across the lifespan underscore the importance of considering sex as a critical factor in neuroscience research. From neurodevelopment to aging, these differences influence the risk, onset, and progression of various neurological and psychiatric disorders. By targeting these sex-specific pathways, future research can develop tailored interventions to support brain health and mitigate disease in both men and women (Table 2).

3 Implications for Mental Health and Disease

3.1 Neurodevelopmental Disorders

Neurodevelopmental disorders such as ASD and ADHD exhibit pronounced sex differences in prevalence, symptomatology, and underlying neurobiological mechanisms [23, 53, 54, 59–61]. A study of over 2,400 individuals with ASD found that females tend to show greater impairments in social communication, overall IQ, and adaptive functioning, whereas males more frequently exhibit restricted and repetitive behaviors.

Males with ASD exhibit enlarged volumes in gray matter, white matter, and the hippocampus, potentially contributing to the overconnectivity observed in certain

Table 2 Sex-specific microglial characteristics across the lifespan and disease contexts

Life stage/context	Male microglia	Female microglia	Implications
Prenatal and early postnatal [8, 26]	↑ Motility, ↑ pro-inflammatory gene expression	↑ Phagocytic activity, ↑ regulatory capacity	Males: ↑ risk of ASD, ADHD due to excessive pruning and inflammation
Adulthood [8, 26]	Persistent inflammatory phenotype (↑ MHC, cytokines)	↑ Reparative function, estrogen-enhanced anti-inflammatory activity	Males: ↑ risk for neuroinflammatory damage Females: Protective effect against MS, Parkinson's
Aging [8, 49]	Continued inflammation, ↓ phagocytosis, impaired debris clearance	Post-menopause: shift to pro-inflammatory state, ↓ hormonal protection	Males: ↑ chronic inflammation and neurodegeneration Females: ↑ Alzheimer's risk and progression
Genetic modulation (e.g. APOE4) [8, 49]	Moderate inflammatory response	↑ Inflammatory response to APOE4, ↑ tau and amyloid pathology	APOE4 more detrimental in females; accelerates Alzheimer's pathology

ASD autism spectrum disorder, *ADHD* Attention deficit hyperactivity disorder, *MHC* major histocompatibility complex, *MS* multiple sclerosis, *APOE4* apolipoprotein E4

brain regions [59, 61]. Conversely, females with ASD show reduced hippocampal volumes, which may explain differences in cognitive and emotional processing. Functional connectivity differences are also evident. Males tend to exhibit hyperconnectivity within local neural networks and hypoconnectivity between long-range networks, while females show more diffuse disruptions that may underlie distinct behavioral profiles.

Estrogen plays a protective role in females by regulating the synthesis and receptor expression of critical neurotransmitters such as GABA and glutamate [42, 62, 63]. For instance, estrogen enhances NMDA receptor expression, supporting synaptic plasticity, while progesterone inhibits excessive glutamatergic responses, which may mitigate excitotoxicity in females with ASD [63]. Males exhibit higher prenatal testosterone exposure, which has been linked to atypical neural development and increased susceptibility to ASD [18, 54, 64]. Testosterone's effects on synaptic pruning and cortical thickness may lower the threshold for ASD-related developmental trajectories in males.

ASD risk genes, such as those involved in synaptic organization (e.g., *SHANK3*, *NRXN1*), show sex-differentiated expression patterns [25, 65, 66]. The X chromosome, which contains genes that escape inactivation in females, provides additional neuroprotective mechanisms absent in males. This difference may partly explain the male bias in ASD prevalence.

Sex differences in ASD suggest the need for tailored diagnostic tools and therapeutic approaches [18, 54]. For example, females with ASD often mask their symptoms, leading to delayed or missed diagnoses. Early interventions should account for these subtleties, emphasizing support for social communication in females and managing behavioral rigidity in males.

ADHD is also more commonly diagnosed in males. Males are more likely to exhibit hyperactive and impulsive behaviors, while females more frequently present with inattentiveness, leading to differential patterns of diagnosis and treatment [19, 46, 67]. Males with ADHD display reduced cortical thickness and smaller volumes in regions such as the prefrontal cortex, which governs attention and impulse control. Females, by contrast, show less pronounced structural deficits but exhibit altered connectivity in networks associated with executive function and emotional regulation. Dopaminergic and noradrenergic signaling, both critical for attention and cognitive control, differ between sexes [19, 34, 68]. Males with ADHD demonstrate greater dopamine receptor dysregulation in the striatum, leading to impulsive behaviors, while females exhibit more subtle deficits in prefrontal dopamine signaling, which may account for their inattentive presentations.

Estrogen's modulation of dopamine and norepinephrine pathways may provide a partial protective effect for females, delaying symptom onset or reducing severity [30, 42, 62, 63]. This hormonal influence may also contribute to the observed shift in ADHD symptoms as females approach puberty.

Sex-specific differences in ADHD call for differentiated treatment approaches. For males, managing hyperactivity and impulsivity through behavioral therapies and stimulant medications may be prioritized. For females, interventions targeting executive functioning and emotional regulation may be more effective.

3.2 *Mood Disorders*

Mood disorders, including depression, anxiety, and post-traumatic stress disorder (PTSD), are among the most common mental health conditions and exhibit significant sex differences in prevalence, symptomatology, and neurobiological mechanisms [31, 45, 69–71]. Women are nearly twice as likely as men to experience depression and anxiety, while men demonstrate greater susceptibility to certain stress-related disorders, such as PTSD. These differences are shaped by hormonal influences, brain circuitry variations, and sex-specific molecular mechanisms, necessitating tailored approaches to diagnosis and treatment.

Women experience major depressive disorder and generalized anxiety disorder at significantly higher rates than men, with a female-to-male ratio of approximately 2:1. These differences are evident in brain structure, function, and neurotransmitter activity. Women with depression show reductions in hippocampal and amygdala volumes, brain regions critical for emotional regulation and memory [71]. In contrast, men exhibit more pronounced structural changes in the anterior cingulate cortex and prefrontal cortex, areas involved in cognitive control and decision-making. Functional MRI studies reveal that corticotropin-releasing factor, a key regulator of the stress response, activates the arousal system more robustly in women than men [71, 72]. This heightened activation may explain the greater sensitivity of women to stress-related mood disturbances.

In depression, serotonin and glutamate signaling exhibit sex-specific alterations [36]. Women show higher expression of serotonin 1D autoreceptors in the dorsal raphe nucleus, which may reduce serotonin availability in the synaptic cleft. Men, by contrast, display more pronounced glutamate-related changes in the prefrontal cortex, affecting cognitive processes such as attention and working memory. GABAergic signaling, essential for inhibitory control and emotional regulation, also differs. Women show reduced GABAergic markers in the basolateral amygdala, while men exhibit similar reductions in the anterior cingulate cortex. These distinctions contribute to variations in anxiety and emotional processing between sexes.

Estrogen and progesterone play pivotal roles in regulating mood [31, 38, 41]. Estrogen enhances serotonergic activity and synaptic plasticity, providing protective effects against mood disorders in premenopausal women. Conversely, hormonal fluctuations during the menstrual cycle, pregnancy, and postpartum periods can destabilize mood regulation, increasing the risk of depressive episodes.

Women often exhibit more affective symptoms, such as sadness, guilt, and fatigue, while men may display more somatic complaints and irritability. Tailoring treatment to these symptom profiles, such as focusing on emotional regulation for women and cognitive therapies for men, can improve outcomes.

PTSD exhibits unique sex differences in molecular and neurocircuitry mechanisms that influence prevalence and clinical presentation [40]. Women are more likely to develop PTSD following trauma, while men are more prone to aggression-related symptoms.

Women with PTSD show dysregulated GABAergic signaling in brain regions such as the prefrontal cortex and basolateral amygdala, impairing the inhibitory control of stress responses [40, 73]. In contrast, men exhibit heightened immune responses and hyperactivation of the hypothalamic–pituitary–adrenal (HPA) axis, exacerbating stress-related arousal. Neuroplasticity-related pathways, such as those involving brain-derived neurotrophic factor, are more disrupted in women, contributing to altered fear extinction and prolonged stress responses [40].

Genomic studies identify differential expression of stress-regulatory genes such as *FKBP5* and *NR3C1*, with men showing greater immune-specific pathway activation and women exhibiting alterations in synaptic plasticity-related genes [29, 37, 40, 41].

Estrogen modulates HPA axis function and enhances fear extinction, which may explain women's heightened vulnerability to PTSD under conditions of hormonal dysregulation, such as menopause [30, 41].

Women with PTSD more frequently report intrusive memories, hyperarousal, and emotional numbing, while men are more likely to exhibit externalizing behaviors such as anger and aggression [40]. These differences suggest the need for sex-specific therapeutic approaches, such as trauma-focused cognitive-behavioral therapy emphasizing emotional processing for women and anger management for men.

3.3 Neurodegenerative Disorders

Neurodegenerative disorders, characterized by progressive neuronal loss and impaired brain function, exhibit significant sex differences in prevalence, progression, and clinical outcomes [19]. Alzheimer's disease is the most common neurodegenerative disorder, with women comprising two-thirds of all cases. In addition to higher prevalence, women also experience faster cognitive decline and greater disease severity compared to men [3]. Estrogen plays a neuroprotective role by enhancing synaptic plasticity, promoting antioxidant defenses, and reducing amyloid beta (Aβ) accumulation [30]. However, the sharp decline in estrogen levels post-menopause increases women's vulnerability to Alzheimer's disease. Estrogen withdrawal accelerates tau hyperphosphorylation and impairs neural repair mechanisms, exacerbating neurodegeneration [19, 30, 74]. Women exhibit greater amyloid plaque deposition and tau pathology than men, even at equivalent stages of cognitive decline. This sex-specific pattern is partly influenced by hormonal shifts and genetic predispositions such as the *APOE4* allele, which disproportionately affects women by amplifying Aβ aggregation and impairing clearance mechanisms. Women with Alzheimer's disease demonstrate faster atrophy in brain regions associated with memory, such as the hippocampus and entorhinal cortex. Men with Alzheimer's disease tend to exhibit more widespread cortical atrophy, affecting multiple functional networks. Depression and sleep disturbances, which are more prevalent in women, further increase Alzheimer's disease risk. Sleep disruptions promote Aβ accumulation, while depression accelerates hippocampal atrophy, compounding cognitive decline. Hormone replacement therapy may offer protective benefits against Alzheimer's disease in postmenopausal women, but its timing and long-term effects require further investigation.

Parkinson's disease, the second most common neurodegenerative disorder, affects men nearly twice as often as women [8, 19, 75, 76]. Despite this, women with Parkinson's disease experience faster disease progression and worse non-motor symptoms. Parkinson's disease is characterized by the loss of dopaminergic neurons in the substantia nigra, leading to motor impairments such as bradykinesia, rigidity, and tremors. Men exhibit more severe dopaminergic loss, contributing to their higher prevalence of motor symptoms [75]. Brain imaging studies reveal greater atrophy in the precentral and postcentral gyri, thalamus, and caudate in men with Parkinson's disease compared to women. This cortical thinning correlates with longer disease duration and older age in men. Women show less severe motor deficits but exhibit greater limbic system involvement, potentially explaining their higher prevalence of anxiety, depression, and sleep disturbances. Estrogen may delay Parkinson's disease onset in women by modulating dopaminergic activity and reducing oxidative stress [30, 38]. However, its protective effects diminish after menopause, accelerating disease progression. Women with Parkinson's disease are more likely to experience non-motor symptoms, including depression, fatigue, and sleep disorders, which negatively impact quality of life. Men, on the other hand, have more pronounced motor impairments and a higher prevalence of rapid eye movement sleep behavior disorder, which is a strong predictor of Parkinson's disease development [76]. Sex-specific treatments, such as estrogen supplementation for women or targeted therapies for rapid eye movement sleep behavior disorder in men, could improve outcomes and slow disease progression. Sex differences in neurodegenerative disorders such as Alzheimer's disease and Parkinson's disease highlight the critical role of hormonal regulation, genetic predispositions, and brain structure in disease pathogenesis and progression. Women's greater susceptibility to Alzheimer's disease is driven by estrogen loss and heightened amyloid and tau pathology, while men's higher Parkinson's disease prevalence reflects more severe dopaminergic degeneration and cortical atrophy. Recognizing these distinctions is vital for developing sex-specific interventions, including hormonal therapies, personalized diagnostics, and tailored treatment strategies to address the unique needs of men and women.

4 Conclusion

Understanding sex differences in neuroscience is essential for a comprehensive view of brain function, development, and disease. From structural and functional divergences in brain anatomy to microglial behavior and neurochemical regulation, accumulating evidence underscores that male and female brains follow distinct biological trajectories shaped by genetic, hormonal, and environmental influences.

This review has highlighted the multilayered nature of sex-specific variation across the nervous system. Anatomical distinctions set the foundation for differential connectivity and functional dynamics, which in turn are modulated by sex-dependent microglial responses and neurotransmitter signaling. These biological differences translate into clear disparities in susceptibility, progression, and outcomes of neurological and psychiatric disorders, such as ASD, Alzheimer's disease, Parkinson's disease, and multiple sclerosis. Importantly, microglial sex differences, shaped by both developmental timing and hormonal milieu, play a central role in these disease patterns, revealing mechanisms that have too often been overlooked in male-biased models of research.

The implications of these findings are both scientific and clinical. They call for the consistent inclusion of SABV in basic and translational neuroscience research. Recognizing sex-specific mechanisms not only enriches our fundamental understanding of brain physiology but also supports the development of precision therapies tailored to the biological context of each individual. Moreover, clinical trials must be designed with sufficient sex-based representation to ensure that therapeutic efficacy and safety are valid across populations.

Future research must deepen our understanding of how sex interacts with genetic factors, epigenetic regulation, and environmental exposures to shape the nervous system over the lifespan. Interdisciplinary approaches—integrating neuroimmunology, endocrinology, systems biology, and computational modeling—are vital to uncovering the full complexity of sex-based neurobiological variation.

Emerging tools and technologies, such as AI, are revolutionizing neuroscience by uncovering subtle sex-specific patterns in brain structure, molecular pathways, and disease progression. AI models trained on large datasets can predict sex-differentiated responses to neuroinflammatory triggers, enhancing our understanding of conditions like Alzheimer's disease and autism. These innovations hold the potential to personalize diagnostics and treatments, addressing the unique needs of men and women.

Despite these advancements, significant challenges remain. Women and other underrepresented groups continue to be excluded from many clinical and preclinical studies, reducing the reliability of research outcomes. Additionally, the lack of standardized methodologies for collecting and analyzing sex-specific data hinders progress. Addressing these barriers requires targeted research funding, equitable representation in studies, and a global commitment to integrating SABV.

Promising initiatives, such as the ALBA Network and the Women's Brain Project, are paving the way for more inclusive and impactful neuroscience research. By fostering diversity, standardizing methodologies, and leveraging tools like post-authorization data collection, the field can generate robust data to inform sex-specific treatments.

Collaboration is essential. Scientists, policymakers, industry leaders, and the public must work together to advance neuroscience research that accounts for sex differences. By prioritizing innovation and equity, we can transform the diagnosis, treatment, and prevention of neurological disorders, ensuring that healthcare solutions meet the diverse needs of men and women alike.

Ultimately, incorporating sex differences into neuroscience is not merely a matter of completeness but a fundamental imperative for scientific accuracy and medical effectiveness. A nuanced understanding of these differences will enable more precise diagnostics, more effective treatments, and more equitable healthcare outcomes. With collective commitment and continued innovation, neuroscience can evolve toward a more inclusive and impactful future.

Acknowledgements Korea Center for Gendered Innovations for Science and Technology research (GISTeR), through the Center for Women in Science, Engineering and Technology (WISET) funded by the Ministry of Science and ICT (WISET202403GI01).

References

1. Becker JB, Prendergast BJ, Liang JW. Female rats are not more variable than male rats: a meta-analysis of neuroscience studies. Biol Sex Differ. 2016;7:34.
2. Rechlin RK, Splinter TFL, Hodges TE, Albert AY, Galea LAM. An analysis of neuroscience and psychiatry papers published from 2009 and 2019 outlines opportunities for increasing discovery of sex differences. Nat Commun. 2022;13:2137.
3. Kim H. Current status and significance of research on sex differences in neuroscience: a narrative review and bibliometric analysis. Ewha Med J. 2024;47: e16.
4. Kiesow H, Dunbar RIM, Kable JW, Kalenscher T, Vogeley K, Schilbach L, et al. 10,000 social brains: sex differentiation in human brain anatomy. Sci Adv. 2020;6:eaaz1170.
5. Xin J, Zhang Y, Tang Y, Yang Y. Brain differences between men and women: evidence from deep learning. Front Neurosci. 2019;13:185.

6. Gualtierotti R, Bressi C, Garavaglia B, Brambilla P. Exploring the impact of sex and gender in brain function: implications and considerations. Adv Ther. 2024;41:4377–83.
7. Shansky RM. Behavioral neuroscience's inevitable SABV growing pains. Trends Neurosci. 2024;47:669–76.
8. Lynch MA. Exploring sex-related differences in microglia may be a game-changer in precision medicine. Front Aging Neurosci. 2022;14: 868448.
9. Casaletto KB, Nichols E, Aslanyan V, Simone SM, Rabin JS, La Joie R, et al. Sex-specific effects of microglial activation on Alzheimer's disease proteinopathy in older adults. Brain. 2022;145:3536–45.
10. Trainor BC, Falkner AL. Quantifying sex differences in behavior in the era of "Big" data. Cold Spring Harb Perspect Biol. 2022;14: a039164.
11. Grabowska A. Sex on the brain: are gender-dependent structural and functional differences associated with behavior? J Neurosci Res. 2017;95:200–12.
12. Ingalhalikar M, Smith A, Parker D, Satterthwaite TD, Elliott MA, Ruparel K, et al. Sex differences in the structural connectome of the human brain. Proc Natl Acad Sci U S A. 2014;111:823–8.
13. Luckhoff HK, Smit R, Phahladira L, du P, Emsley R, Asmal L. Sex versus gender associations with brain structure. J Clin Neurosci. 2024;122:103–9.
14. Seney ML, Glausier J, Sibille E. Large-scale transcriptomics studies provide insight into sex differences in depression. Biol Psychiatry. 2022;91:14–24.
15. Zhang K, Fang H, Li Z, Ren T, Li B-m, Wang C. Sex differences in large-scale brain network connectivity for mental rotation performance. NeuroImage. 2024;298:120807.
16. Luders E, Gaser C, Narr KL, Toga AW. Why sex matters: brain size independent differences in gray matter distributions between men and women. J Neurosci. 2009;29:14265–70.
17. Ferretti MT, Iulita MF, Cavedo E, Chiesa PA, Schumacher Dimech A, Santuccione Chadha A, et al. Sex differences in Alzheimer disease—the gateway to precision medicine. Nat Rev Neurol. 2018;14:457–69.
18. Ferri SL, Abel T, Brodkin ES. Sex differences in autism spectrum disorder: a review. Curr Psychiatry Rep. 2018;20:1–17.
19. Pinares-Garcia P, Stratikopoulos M, Zagato A, Loke H, Lee J. Sex: a significant risk factor for neurodevelopmental and neurodegenerative disorders. Brain Sci. 2018;8:154.
20. Loosen AM, Kato A, Gu X. Revisiting the role of computational neuroimaging in the era of integrative neuroscience. Neuropsychopharmacology. 2024;50:103–13.
21. Urai AE, Doiron B, Leifer AM, Churchland AK. Large-scale neural recordings call for new insights to link brain and behavior. Nat Neurosci. 2022;25:11–9.
22. Verriotis M, Jones L, Whitehead K, Laudiano-Dray M, Panayotidis I, Patel H, et al. The distribution of pain activity across the human neonatal brain is sex dependent. Neuroimage. 2018;178:69–77.
23. Guang S, Pang N, Deng X, Yang L, He F, Wu L, et al. Synaptopathology involved in autism spectrum disorder. Front Cell Neurosci. 2018;12:470.
24. Chung L, Bey AL, Jiang YH. Synaptic plasticity in mouse models of autism spectrum disorders. Korean J Physiol Pharmacol. 2012;16:369–78.
25. Mottron L, Duret P, Mueller S, Moore RD, Forgeot d'Arc B, Jacquemont S, et al. Sex differences in brain plasticity: a new hypothesis for sex ratio bias in autism. Mol Autism. 2015;6:33.
26. Pramanik S, Devi MH, Chakrabarty S, Paylar B, Pradhan A, Thaker M, et al. Microglia signaling in health and disease—implications in sex-specific brain development and plasticity. Neurosci Biobehav Rev. 2024;165: 105834.
27. Bobotis BC, Braniff O, Gargus M, Akinluyi ET, Awogbindin IO, Tremblay M-È. Sex differences of microglia in the healthy brain from embryonic development to adulthood and across lifestyle influences. Brain Res Bull. 2023;202: 110752.
28. Neale KJ, Reid HMO, Sousa B, McDonagh E, Morrison J, Shultz S, et al. Repeated mild traumatic brain injury causes sex-specific increases in cell proliferation and inflammation in juvenile rats. J Neuroinflammation. 2023;20:250.
29. Wellman CL, Bangasser DA, Bollinger JL, Coutellier L, Logrip ML, Moench KM, et al. Sex differences in risk and resilience: stress effects on the neural substrates of emotion and motivation. J Neurosci. 2018;38:9423–32.

30. Wise PM, Dubal DB, Wilson ME, Rau SW, Böttner M. Minireview: Neuroprotective effects of estrogen—new insights into mechanisms of action. Endocrinology. 2001;142:969–73.
31. Hyer MM, Phillips LL, Neigh GN. Sex differences in synaptic plasticity: hormones and beyond. Front Mol Neurosci. 2018;11:266.
32. Tavares-Ferreira D, Ray PR, Sankaranarayanan I, Mejia GL, Wangzhou A, Shiers S, et al. Sex differences in nociceptor translatomes contribute to divergent prostaglandin signaling in male and female mice. Biol Psychiatry. 2022;91:129–40.
33. Jaeckel ER, Herrera YN, Schulz S, Birdsong WT. Chronic morphine induces adaptations in opioid receptor signaling in a thalamostriatal circuit that are location dependent, sex specific, and regulated by μ-opioid receptor phosphorylation. J Neurosci. 2024;44: e0293232023.
34. Costa KM, Schenkel D, Roeper J. Sex-dependent alterations in behavior, drug responses and dopamine transporter expression in heterozygous DAT-Cre mice. Sci Rep. 2021;11:3334.
35. Kalamarides DJ, Singh A, Wolfman SL, Dani JA. Sex differences in VTA GABA transmission and plasticity during opioid withdrawal. Sci Rep. 2023;13:8460.
36. Jones MD, Lucki I. Sex differences in the regulation of serotonergic transmission and behavior in 5-HT receptor knockout mice. Neuropsychopharmacology. 2005;30:1039–47.
37. Luo Y-J, Bao H, Crowther A, Li Y-D, Chen Z-K, Tart DS, et al. Sex-specific expression of distinct serotonin receptors mediates stress vulnerability of adult hippocampal neural stem cells in mice. Cell Rep. 2024;43: 114140.
38. Bendis PC, Zimmerman S, Onisiforou A, Zanos P, Georgiou P. The impact of estradiol on serotonin, glutamate, and dopamine systems. Front Neurosci. 2024;18.
39. Chavez C, Hollaus M, Scarr E, Pavey G, Gogos A, van den Buuse M. The effect of estrogen on dopamine and serotonin receptor and transporter levels in the brain: an autoradiography study. Brain Res. 2010;1321:51–9.
40. Wang J, Zhao H, Girgenti MJ. Posttraumatic stress disorder brain transcriptomics: convergent genomic signatures across biological sex. Biol Psychiatry. 2022;91:6–13.
41. Tongta S, Daendee S, Kalandakanond-Thongsong S. Effects of estrogen receptor β or G protein-coupled receptor 30 activation on anxiety-like behaviors in relation to GABAergic transmission in stress-ovariectomized rats. Neurosci Lett. 2022;789: 136885.
42. Herbison AE. Estrogen regulation of GABA transmission in rat preoptic area. Brain Res Bull. 1997;44:321–6.
43. Averitt DL, Hornung RS, Murphy AZ. Role of sex hormones on pain. Neuroscience. 2019.
44. Mogil JS. Sex differences in pain and pain inhibition: multiple explanations of a controversial phenomenon. Nat Rev Neurosci. 2012;13:859–66.
45. Seney ML, Sibille E. Sex differences in mood disorders: perspectives from humans and rodent models. Biol Sex Differ. 2014;5:17.
46. Arnett AB, Pennington BF, Willcutt EG, DeFries JC, Olson RK. Sex differences in ADHD symptom severity. J Child Psychol Psychiatry. 2015;56:632–9.
47. Prinz M, Jung S, Priller J. Microglia biology: one century of evolving concepts. Cell. 2019;179:292–311.
48. Paolicelli RC, Sierra A, Stevens B, Tremblay M-E, Aguzzi A, Ajami B, et al. Microglia states and nomenclature: a field at its crossroads. Neuron. 2022;110:3458–83.
49. Yanguas-Casás N. Physiological sex differences in microglia and their relevance in neurological disorders. Neuroimmunol Neuroinflammation. 2020;7:13–22.
50. Villa A, Gelosa P, Castiglioni L, Cimino M, Rizzi N, Pepe G, et al. Sex-specific features of microglia from adult mice. Cell Rep. 2018;23:3501–11.
51. Guneykaya D, Ivanov A, Hernandez DP, Haage V, Wojtas B, Meyer N, et al. Transcriptional and translational differences of microglia from male and female brains. Cell Rep. 2018;24:2773-83.e6.
52. Han J, Fan Y, Zhou K, Blomgren K, Harris RA. Uncovering sex differences of rodent microglia. J Neuroinflammation. 2021;18:74.
53. Granerud G, Elvsåshagen T, Arntzen E, Juhasz K, Emilsen NM, Sønderby IE, et al. A family study of symbolic learning and synaptic plasticity in autism spectrum disorder. Front Hum Neurosci. 2022;16: 950922.

54. Napolitano A, Schiavi S, La Rosa P, Rossi-Espagnet MC, Petrillo S, Bottino F, et al. Sex differences in autism spectrum disorder: diagnostic, neurobiological, and behavioral features. Front Psychiatry. 2022;13: 889636.
55. McCarthy MM, Pickett LA, VanRyzin JW, Kight KE. Surprising origins of sex differences in the brain. Horm Behav. 2015;76:3–10.
56. Castro-Aldrete L, Moser MV, Putignano G, Ferretti MT, Schumacher Dimech A, Santuccione CA. Sex and gender considerations in Alzheimer's disease: the women's brain project contribution. Front Aging Neurosci. 2023;15:1105620.
57. Arnold AP, Klein SL, McCarthy MM, Mogil JS. Male–female comparisons are powerful in biomedical research—don't abandon them. Nature. 2024;629:37–40.
58. Rosmus DD, Lange C, Ludwig F, Ajami B, Wieghofer P. The role of osteopontin in microglia biology: current concepts and future perspectives. Biomedicines. 2022;10:840.
59. Hazlett HC, Gu H, Munsell BC, Kim SH, Styner M, Wolff JJ, et al. Early brain development in infants at high risk for autism spectrum disorder. Nature. 2017;542:348–51.
60. Li Y, Li R, Wang N, Gu J, Gao J. Gender effects on autism spectrum disorder: a multi-site resting-state functional magnetic resonance imaging study of transcriptome-neuroimaging. Front Neurosci. 2023;17:1203690.
61. Smith REW, Avery JA, Wallace GL, Kenworthy L, Gotts SJ, Martin A. Sex differences in resting-state functional connectivity of the cerebellum in autism spectrum disorder. Front Hum Neurosci. 2019;13:104.
62. Hoffman EJ, Turner KJ, Fernandez JM, Cifuentes D, Ghosh M, Ijaz S, et al. Estrogens suppress a behavioral phenotype in zebrafish mutants of the autism risk gene, CNTNAP2. Neuron. 2016;89:725–33.
63. Molinaro G, Bowles JE, Croom K, Gonzalez D, Mirjafary S, Birnbaum SG, et al. Female-specific dysfunction of sensory neocortical circuits in a mouse model of autism mediated by mGluR5 and estrogen receptor α. Cell Rep. 2024;43: 114056.
64. Baron-Cohen S. The extreme male brain theory of autism. Trends Cogn Sci. 2002;6:248–54.
65. Braden BB, Wallace GL, Floris DL, Pelphrey KA. Editorial: sex differences in the autistic brain. Front Integr Neurosci. 2021;15: 708368.
66. Mo K, Sadoway T, Bonato S, Ameis SH, Anagnostou E, Lerch JP, et al. Sex/gender differences in the human autistic brains: a systematic review of 20 years of neuroimaging research. Neuroimage Clin. 2021;32: 102811.
67. Babinski DE. Sex differences in ADHD: review and priorities for future research. Curr Psychiatry Rep. 2024;26:151–6.
68. Williams OOF, Coppolino M, George SR, Perreault ML. Sex differences in dopamine receptors and relevance to neuropsychiatric disorders. Brain Sci. 2021;11:1199.
69. Kaluve AM, Le JT, Graham BM. Female rodents are not more variable than male rodents: a meta-analysis of preclinical studies of fear and anxiety. Neurosci Biobehav Rev. 2022;143: 104962.
70. GBD 2019 Mental Disorders Collaborators. Global, regional, and national burden of 12 mental disorders in 204 countries and territories, 1990–2019: a systematic analysis for the Global Burden of Disease Study 2019. Lancet Psychiatry. 2022;9:137–50.
71. Bangasser DA, Cuarenta A. Sex differences in anxiety and depression: circuits and mechanisms. Nat Rev Neurosci. 2021;22:674–84.
72. Kaczkurkin AN, Raznahan A, Satterthwaite TD. Sex differences in the developing brain: insights from multimodal neuroimaging. Neuropsychopharmacology. 2019;44:71–85.
73. Seney ML, Nestler EJ. Introduction to special issue: insight into sex differences in neuropsychiatric syndromes from transcriptomic analyses. Biol Psychiatry. 2022;91:3–5.
74. Klein SL, Flanagan KL. Sex differences in immune responses. Nat Rev Immunol. 2016;16:626–38.
75. Beheshti I, Booth S, Ko JH. Differences in brain aging between sexes in Parkinson's disease. NPJ Parkinsons Dis. 2024;10:35.
76. Gillies GE, Pienaar IS, Vohra S, Qamhawi Z. Sex differences in Parkinson's disease. Front Neuroendocrinol. 2014;35:370–84.

Neuroimaging Insight into Sex and Gender Differences in Brain Health and Disease

Melani Macik, Hyang Woon Lee, and R. Todd Constable

Abstract In neuroimaging and neuroscience research, investigating sex differences is crucial for understanding unique aspects of development, behavior, cognitive processes, symptomology, and disease prognosis within the nervous system. While exploring sex differences in neuroimaging requires rigorous scientific standards and often faces criticism [1, 2], such research is essential to enhance representation in science and improve patient care. Neuroimaging research on sex differences (NISD) has highlighted sex-specific vulnerabilities: for examples, men are more frequently affected by Autism Spectrum Disorder, Parkinson's disease, traumatic brain injury, and strokes, while women are more prone to posttraumatic stress disorder, mood disorders (bipolar/depression), Alzheimer's disease, and multiple sclerosis [3]. Elucidating NISD in these health conditions can uncover distinct mechanisms underlying their development. This chapter reviews recent advances in NISD, focusing on primary research in structural, functional, task-based, and endocrinological methods in neuroimaging. These studies detail the importance of integrating sex differences into neuroscience to better address and treat diverse neuropsychiatric diseases.

M. Macik · R. T. Constable (✉)
Interdepartmental Neuroscience Program, Yale University, New Haven, CT, USA
e-mail: todd.constable@yale.edu

M. Macik
e-mail: melani.macik@yale.edu

M. Macik
Department of Neuroscience, Yale University, New Haven, CT, USA

H. W. Lee
Departments of Neurology and Medical Science, College of Medicine and Ewha Medical Research Institute, Ewha Womans University, Seoul, Republic of Korea

Computational Medicine, System Health Science and Engineering, and Artificial Intelligence Convergence Graduate Programs, Ewha Womans University, Seoul, Republic of Korea

H. W. Lee
e-mail: leeh@ewha.ac.kr

R. T. Constable
Departments of Radiology, Medical Imaging and Biomedical Engineering, Yale University School of Medicine, New Haven, CT, USA

H. Lee et al. (eds.), *Sex, Gender, and Emerging Technology in Healthcare: Mitigating Bias and Fostering Equity*, https://doi.org/10.1007/978-981-95-2070-1_5

Keywords Neuroimaging · Sex · Gender · Brain · Health · Disease

1 Introduction

The study of sex differences in neuroimaging has become a central focus in neuroscience, providing insights into how biological sex impacts brain structure, function, and connectivity. Understanding these differences is crucial for interpreting variations in brain development, cognition, behavior, and vulnerability to neurological and psychiatric disorders. Men and women are affected differently by many neuropsychiatric conditions. Autism spectrum disorder (ASD), Parkinson's disease (PD), Alzheimer's disease (AD), mood and sleep disorders impact men and women differently, suggesting inherent distinctions in brain processes that neuroimaging can help reveal. Investigating these distinctions enables a better understanding of the mechanisms underlying sex-specific patterns of health and disease. However, the full range of these mechanisms underlying these sex and gender differences remains to be uncovered.

Recent advancements in neuroimaging techniques have enhanced our ability to investigate sex differences in the brain with greater precision. Structural imaging, functional MRI, task-based imaging, and hormone-related neuroimaging are some methods that have expanded the field. Combined with tools like machine learning, these techniques have begun to reveal complex interactions between brain anatomy, hormone influences, and functional connectivity (FC) that may vary between sexes. Insights from this research are critical, as it has the potential to inform personalized medicine, potentially enhancing diagnostic accuracy and treatment efficacy by considering sex-based variations in brain function.

Exploring sex and gender differences in neuroimaging also addresses an important gap in representation of scientific research and medical practice. Historically, medical and neuroscience research has underrepresented women, leading to gaps in knowledge and potential biases in treatment approaches. Neuroimaging research focused on sex and gender differences has the potential to close these gaps, potentially improving health outcomes for both men and women.

Hormonal differences represent one of the most significant explanations for sex differences in brain function, especially in disorders that affect men and women differently. Divergent functional patterns often associate with variations in sex hormone levels and their dynamic impact on the brain.

It is important to note that these sex and gender differences are influenced not only by biological mechanisms including hormonal and/or genetic factors, but also by various social, lifestyle and environmental consequences. In other words, brain functions are influenced by various biological, genetic, environmental, and lifestyle factors, which can be intertwined with sex/gender differences to shape a unique pattern in terms of behavior, and/or development of various neuropsychiatric diseases. Consequently, sex- and gender-based variations in brain function are part of a complex, multifaceted issue that demands further extensive exploration.

This chapter highlights the importance of sex- and gender-based neuroimaging research. Neuroimaging research on sex differences (NISD) not only enhances our understanding of brain function but also offers pathways to develop more targeted and effective clinical interventions. As this field continues to evolve, it may reshape our approach to brain health across sexes, ultimately contributing to more inclusive and effective healthcare practices.

2 Sex/Gender Differences in Structural Neuroimaging

Mass-scale studies prominently contribute to advances in structural NISD research, largely due to the accessibility of the Human Connectome Project and other similar databases. One such project created a reproducible spatial map of sex-specific gray matter volume (GMV) by combining structural neuroimaging data from two data sets, for a total of over 2,000 brains. In this map, females show increased GMV in the prefrontal and superior parietal cortices, while male spatial maps show bias in the ventral occipitotemporal and subcortical regions. These researchers also found sex differences in facial processing regions. GMV sex differences were associated with regional expression of sex-chromosome genes [4]. These results indicate a link between sex differences in gene expression and structural development. Another massive project studied sex differences in connections between genetic makeup and neuroanatomy, examining over 1,000 phenotypes. The conductors of this project found high heritability in females for V2 cortical structure and in men higher tissue-specific expression of *RBFOX1*, a gene linked to neuropsychiatric disorders that predominantly affect males [5]. Additionally, a study of 960 healthy individuals found consistent sex differences in the structure of the hippocampus. These differences were most prominent in the parasubiculum and the fimbria, in both of which males were larger [6]. These large-scale studies provide data more representative of the general population, so they can detect pervasive sex differences in neuroanatomy.

Large-scale neuroimaging projects also advance understanding of neurodevelopment and sex differences in pediatric populations. One project developed a machine learning model to classify sex based on morphometry and image intensity from structural MRI data of 8,325 adolescents (ages 9–10 years), achieving an 86% accuracy rate. This model identified the insula, precentral and postcentral gyri, and pericallosal sulcus as key predictive regions, indicating consistent sex-specific structural patterns at a young age [7]. Another study of nearly 1,000 children (ages 6–22 years) used diffusion tensor imaging (DTI) to model structural connectome. Males had higher within-hemispheric connectivity, modularity, and transitivity, while females showed increased between-hemispheric and cross-module connectivity. These findings align with previous notions that male and female brains specialize in different cognitive domains, specifically motor and spatial processing for males and memory and social cognition for females [8]. A study of 371 participants (ages 8–12 years) further highlighted sex differences in the developing cerebellum, revealing that males showed greater GBV in the right lobules I–IV. Higher cognitive function correlated with greater volume in the left lobule VI [9]. These structural variations could underlie sex differences in motor abilities. Overall, structural NISD in pediatric populations indicate neurodevelopmental mechanisms in males and females.

Researchers have also explored NISD in the context of neuropsychiatric disorders, revealing crucial distinctions that could inform treatment. One study on schizophrenia found sex-specific structural differences in brain areas such as the frontomedial cortex, basal forebrain, cingulate and paracingulate gyri, posterior supramarginal gyrus (SMG), and planum temporale. These regions are known to be sexually dimorphic in healthy populations and affected by schizophrenia [10]. Similarly, in adolescents (ages 13–21 years) with bipolar disorder, sex-based volumetric differences were observed. Typically, healthy females show lower volumes in the SMG and higher volumes in the inferior parietal lobe (IPL). In bipolar disorder, however, females showed higher volume in SMG and a greater magnitude of increased volume in IPL [11], suggesting a potential reversal in typical sex differences that could explain the pathophysiologic mechanisms of this disorder.

Sex-specific structural differences have been revealed in major depressive disorder (MDD) as well. Females with MDD displayed smaller size in the prefrontal cortex and increased gyrification of the left visual cortex compared to healthy control, while males revealed opposite trends in these regions. Individuals with MDD had increased gyrification in posterior brain regions, regardless of sex [12]. These findings highlight the importance of sex-specific considerations in MDD treatment strategies. Given the relevance of the limbic system in mood disorders, another study investigated NISD in this region, using volumetric imaging data from 1,017 subjects. Limbic structures showed greater sex differences and heritability than non-limbic areas, to the extent that machine learning models could accurately predict sex based on the limbic system data alone and perform comparatively well to whole-brain data. Although limbic structures were associated with MDD diagnoses, this association diminished after adjusting for the total intracranial volume, highlighting the complex relationship between sex, limbic structure, and mood disorders [13]. These findings provide further evidence for the structural basis for sex differences in neuropsychiatric conditions and emphasize the need for sex-specific approaches in clinical practice in MDD.

In addition, NISD research provides interesting insights into various domains, including addiction, neurodegenerative disorders, early life adversity (ELA), and physiologic brain aging, revealing both structural and functional sex-specific patterns across the lifespan. For addiction, NISD studies have identified sex-based structural differences associated with substance use. One study used T1-weighted MRI to investigate the impact of alcohol use disorder (AUD) on neuroanatomy. The results reported significant volumetric atrophy in the amygdala and its subregions in males with AUD when compared to control males. However, AUD cases of both sexes display smaller volumes in the hippocampus and its subregions [14]. In nicotine addiction, which affects women more than men, researchers investigated cigarette craving and the insula cortex structure before and after smoking the first cigarette of the day. Although there was no sex difference in the positive association between nicotine dependence and craving, women reported stronger craving than men at both time points. Only in women was craving negatively associated with the cortical thickness of the right anterior insular sulcus [15]. These studies illustrate the importance of examining sex-specific neurobiological mechanisms in addiction.

NISD research has also made substantial contributions to understanding neurodegenerative disorders, such as PD and AD. In PD, T1-weighted and diffusion neuroimaging revealed that male patients experience extensive atrophy and disrupted connectivity compared to females in 11 regions, including the bilateral frontal lobes, left insular lobe, and right postcentral gyrus. In contrast, females displayed more tissue loss in 6 regions, such as the left frontal, right parietal, and right occipital cortices. While these findings are informative of sex-specific mechanisms of neurodegeneration in PD, there was a notable imbalance in the ratio of women to men (149M/83F for PD and 78M/39F for healthy controls) [16]. Another study developed a sex-specific diagnostic machine learning model for PD that performs more accurately than models that combine both sexes [17]. In AD, PET and MRI analyses have shown that male patients experience accelerated Tau protein accumulation and brain volume declines when compared to females [2]. Additionally, associations between inflammatory biomarkers, such as interleukin (IL)-6 and IL-8, and total cerebral brain volume were observed in males but not females with AD [18]. These findings support sex-specific mechanisms for the development of neurodegeneration, which could have implications for individualized treatment approaches.

Sex differences in structural development have also provided crucial information in the context of ELA [19]. ELA has been established to stunt central nervous system (CNS) development and create susceptibility to cognitive, immune, and other functional challenges [19]. One study found, along with greater motor and cognitive impairments, males showed decreased white matter functional anisotropy across development (ages 24 weeks preterm to 8 years) following early life brain injury [20]. Another study showed that high childhood stress associated with long-term microstructural changes in the hippocampus, amygdala, and thalamus, regardless of sex [21]. Additionally, sex-specific differences appear in structural biomarkers for resilience, or the ability to recover from adversity, in youth (ages 9–18 years). Resilience was linked to higher GMV in the right inferior and medial frontal gyri in both sexes. Separately, males showed resilience-related GMV increases in the left middle temporal gyrus, while females showed a negative association between resilience and GMV in this region. Additionally, females displayed positive associations with resilience and GMV of the right middle frontal gyrus and bilateral cerebellar tonsils, contrasting with a negative association in males. These results indicate some unified and some divergent mechanisms for resilience in youth [22]. Another study revealed a positive association between resilience and GMV in the ventrolateral prefrontal cortex projecting to the anterior insula. Males scored significantly higher than females both in this association and in psychological resilience measures alone [23]. These findings indicate that NISD research on resilience can provide valuable insights for understanding differential long-term outcomes following ELA.

In aging research, structural MRI of older adults (ages 65 years and older) revealed that ELA was associated with reduced volume in the frontal, cingulate, and parietal cortices, while regions including the amygdala, caudate, putamen, pallidum, and nucleus accumbens (NAc) showed increased volumes [24]. These areas are implicated in schizophrenia and PD, indicating that ELA may increase vulnerability to these neuropsychiatric disorders later in life. Research on healthy aging also highlights sex-based patterns: men lose more volume in the frontal and temporal lobes, while women experience greater volume loss in the hippocampus and parietal lobes. Metabolically, while men showed less asymmetry in aging, women had greater metabolic reductions in the hippocampus and thalamus [25]. Additionally, a brain aging study of 2,640 healthy participants revealed sex differences in age-related sleep deterioration and sleep-related brain aging patterns. Women showed greater declines in sleep quality, duration, and efficiencies compared to men, even before the age 60, which was related to age-related cognitive declines and reduction in brain volumes [26, 27].

Studies integrating structural imaging with FC analyses have deepened our understanding of how anatomy and function interact across sex. For instance, a study examining GMV and FC in 290 healthy adults reported that females exhibited greater GMV and FC values in the frontal, parietal, occipital, and cerebellar regions, but no relationship between structural and functional changes [28]. Another study examined loneliness and its neurobiological correlates, identifying the default mode network (DMN) as strongly associated with perceived social isolation in both sexes, but somatomotor network associations to loneliness were much stronger in men than women [29]. These data show possible neurobiological underpinnings for sex-specific presentations of loneliness. Lastly, research on the functional organization of the sensory-association pathways in 1000 healthy brains found sex differences that were not related to brain size or microstructure, but instead to FC and network topology [30]. These studies emphasize the value of integrating structural and functional NISD research, providing a more comprehensive picture of how sex-specific neurobiological processes shape development, resilience and vulnerability to various neuropsychiatric diseases.

3 Sex/Gender Differences in Functional Neuroimaging

3.1 Connectivity-Based Functional Neuroimaging

Resting-state functional connectivity (rs-FC) has garnered attention in NISD research fields. A study examined signal strength through sample entropy in resting-state functional MRI (rs-fMRI) of 1,642 healthy individuals. Females exhibited higher signal complexity across all functional brain networks in compared to males, with the most significant sex differences observed in DMN. Both sexes showed relatively high complexity in the frontoparietal control and cingulo-opercular networks (CON), but low complexity in the cerebellum and sensorimotor networks [31]. Another large-scale project assessed associations between aggression and rs-FC data from 745 individuals. Significant rs-FC was identified in the middle temporal and occipital gyri across all participants, and both regions associated with reduced aggression. Additional significant rs-FC in the thalamus, and a negative correlation between rs-FC in the lateral hypothalamus and aggression, were discovered in males. In contrast, females revealed no significant rs-FC clusters in the lateral or medial hypothalamus [32]. Further investigations into the hippocampus highlight sex differences in FC. A study using rs-FC and T2-weighted fMRI in 32 healthy adults found that males had stronger interhemispheric FC between the left and right hippocampus in males compared to females [33]. Another project from the same group investigated participants with mild cognitive impairment (MCI), the prodromal stage for AD. When compared to males, female MCI phenotypes exhibited more severe cognitive impairment, as well as increased hippocampal connectivity to the precuneus cortex and brainstem [34]. These two studies provide potential sex-specific mechanisms in the progression of AD.

Functional connectivity NISD (FC NISD) can be very informative for typical neurodevelopment, given the different maturation rates of male and female brains. One study on fetal FC explored associations with gestational age (GA). Females exhibited stronger FC-GA associations compared to males in connections with the posterior cingulate cortex (PCC) to the left temporal pole, the cerebellum to the left superior frontal gyrus, the left visual cortex to the subcortical region, and the prefrontal cortex to the subcortical and temporal regions. Conversely, males showed greater FC-GA associations in the cerebellum [35]. Research into infant resting-state networks revealed positive associations between age and connectivity across four resting-state networks, providing a link between chronological and neural progress. The only sex difference occurred in the inferotemporal regions of the visual association network, in which females showed higher connectivity. Importantly, preterm birth was associated with FC impairments across all networks and increased FC in the superior parietal lobules [36].

A longitudinal study analyzed the developmental trajectory of rs-FC in the DMN of children (ages 6–12 years) and young adults (ages 18–28 years), observing increased complexity with age. DMN connectivity in late childhood spatially aligned with that of the young adults with no significant sex differences or interactions, revealing consistent DMN maturation across both sexes [37]. These findings collectively emphasize how sex differences in resting-state connectivity can shed light on unique neurodevelopmental pathways and potential risk factors for neuropsychiatric and neurodegenerative conditions.

ASD, a neurodevelopmental disorder characterized by various cognitive, motor, and social symptoms, has been tied to disrupted dynamic functional networks. Since males and females often present ASD differently, investigating sex differences in functional networks of affected individuals offers valuable insight. One research team explored multilayer network analysis and dynamic network reconfiguration in the context of ASD. They discovered that females with ASD displayed lower network flexibility than healthy counterparts in several areas, including the cingulo-opercular (CON), central executive (CEN), salience (SEN), and subcortical (SUB) networks. Additionally, males with ASD had higher flexibility scores than ASD females in the SEN and SUB. Notably, reduced flexibility in females with ASD predict symptom severity, especially in social communication deficits, stereotyped behaviors, and restricted interests [38]. Another study of functional brain connectivity in ASD revealed increased connectivity in the default mode and executive control networks across both sexes. Among patients, males showed particularly elevated connectivity in the cerebellum. ASD females displayed connectivity alterations between visual, language, and basal ganglia networks in comparison to neurotypical females [39]. A related study discovered greater FC in females than males with ASD in the default mode and CEN. Interestingly, the only significant sex difference in healthy controls occurred in SEN, in which females show greater connectivity [40]. Advancements of NISD research highlight potentially divergent mechanisms in the development of ASD in males and females.

Researchers have also analyzed rs-FC in the amygdala in conjunction with psychiatric symptoms in adolescents (ages 8–17 years). Internalizing symptoms often increase severity in anxiety or depressive disorders. This study discovered that in males, anxious-depressed symptoms were negatively correlated with connectivity in the subcallosal prefrontal cortex, whereas in females, internalizing, anxious-depressed, and somatic symptoms were positively associated with connectivity in the cingulate gyrus, insula, and somatosensory cortex. In general, internalizing symptoms correlated with increased amygdala rs-FC to networks involved in emotional and somatosensory processing, salience detection, and action selection [41]. These results potentially provide evidence for sex-specific development of anxiety disorders, which often present differently in males and females.

Functional dynamics and NISD have been explored in other neuropsychiatric disorders as well. For instance, a study used DTI and fMRI to understand the onset of posttraumatic stress disorder (PTSD). Females receive diagnoses of PTSD more often than males, so exploring sex differences could provide evidence for developmental mechanisms. This study gathered fMRI data two weeks following a traumatic incident and examined PTSD symptoms after six months. In females, decreased activity in the ventromedial prefrontal cortex at two weeks predicted higher PTSD symptoms at six months. In males, increased activity in the inferior frontal gyrus predicted lower symptom scores [42]. These results suggest sex-specific dynamics involved in the development of PTSD. Similar to PTSD, bipolar disorder is more prevalent in females. A research team used arterial spin labeling perfusion MRI to investigate cerebral blood flow (CBF) in adolescent bipolar patients and healthy controls (ages 13–20 years). Bipolar patients cohesively displayed higher CBF in the superior longitudinal fasciculus. However, female patients showed significantly higher CBF than male patients in the precuneus/ PCC, left frontal and occipital poles, left thalamus, left superior longitudinal fasciculus, and left precentral gyrus. Females also exhibited a sex- and diagnosis-specific increase in CBF in the precuneus/PCC [43]. These results implicate sex-specific connectivity of several regions, especially the precuneus/PCC, in the development of bipolar disorder.

Another study investigated sex differences in FC and ingestive behaviors among individuals with food addiction (FA). Overall, FA correlated with increased connectivity between the brainstem and orbital frontal gyrus, and within the reward network. Importantly, connectivity in the reward network was associated positively with food cravings, implicating this network's potential role in FA severity. Females with FA displayed higher connectivity than males within the salience and emotional regulation networks, which possibly relates to emotional overeating. Conversely, FA males showed higher connectivity in the default mode and central executive networks, which could relate to habitual and comfort-driven behaviors [44]. These studies collectively underscore the importance of investigating sex-specific neurobiological mechanisms in neuropsychiatric disorders, which could inform targeted interventions and treatments.

NISD research also provides insight into neurodegenerative processes. In a study of rs-FC in multiple sclerosis (MS), which affects women more severely, researchers found a wide array of connectivity alterations. Among both sexes, MS associated with increased connectivity from the posterior cingulate to several regions, including the bilateral medial frontal gyri, left ventral anterior cingulate cortex (ACC), right putamen, and left middle temporal cortex. Male MS patients showed weaker connectivity to the caudate than female patients, while female patients and male controls displayed higher rs-FC between the left dorsal lateral prefrontal cortex and the posterior cingulate in comparison with female controls [45]. These results convey divergent neural effects of MS by sex. Conversely to MS, traumatic brain injury (TBI)-related degeneration affects males more severely than females. Research revealed higher cortical myelination in females than males following brain injury, which could explain some sex differences in TBI outcomes. However, age was a stronger predictor of demyelination and cognitive decline in both sexes [46].

A clinical trial investigated sex differences in the prospect of transcranial direct current stimulation (tDCS) as a treatment for Primary Progressive Aphasia (PPA), a neurodegenerative disorder that impairs language functions. Active tDCS modulates FC, and it led treated PPA females to experience increased connectivity in the language network compared to untreated females. Additionally, PPA males with treatment show increased FC in the DMN compared to untreated males. Of note, PPA males show increased FC in the DMN compared to females, regardless of treatment status. FC in language regions decreased as a factor of age across groups [47]. This trial suggests sex-specific mechanisms by which tDCS modulates connectivity in PPA. In total, functional NISD research can provide information on the development and treatment of neurodegenerative disorders.

Computational networks and machine-based learning likely comprise the future of neuroimaging analysis. One study used fMRI and multivariate pattern analysis to develop a machine learning model capable of predicting sex based on brain topology with an accuracy of 82.9%. Key brain regions in association networks such as the ventral attention, default mode, and frontoparietal networks, were the most informative for prediction [48]. Another study developed a machine learning model to classify sex based on rs-FC, achieving a mean accuracy of 68.7%. The cingulate cortex, medial and lateral frontal cortex, temporoparietal regions, insula, and precuneus were the most informative for prediction [49]. These advancements in machine learning models and computational networks will likely pave the way for future NISD research.

3.2 Task-Based Functional Neuroimaging

Task-based NISD research reports how women and men use different parts of their brains when approaching the same task. Several studies have defined mechanistic differences in approach to mental rotation tasks. An fMRI project showed males exhibit less cross-network interaction of the visual network, but more intra-network integration and cross-network interaction of SEN. Importantly, the elevated cross-network interactions for each sex (visual for females, salience for males) mediated any sex differences in performance [50]. Another study involving a mental rotation task found no sex differences in accuracy or performance on tests. However, this project detected higher frontal activation in females and stronger parietal activation in males [51]. A similar project reiterated these findings [52]. One group combined motor, language, and emotional tests in their study of task-based functional neuroimaging. The results report in males faster reaction times for motor tasks, and more accuracy for language and spatial reasoning tasks. Faster reaction times for emotional tasks and nonverbal reasoning were observed in females. As far as connectivity goes, females displayed stronger within-nodule connectivity, while between-nodule connectivity was higher in males [53].

Task-based NISD research often examines memory acuity as well. One study administered a working memory task and conducted MRI scans on elderly (mean 77 years) participants. Results report a positive correlation between precuneus volumes and task performance in both sexes. The left precuneus was more heavily associated with accuracy and reaction time in females. Men displayed faster reaction times and women were more accurate overall [54]). Another project examined FC during an episodic memory task across age groups (19–76 years). Elevated activity in the dorsal attention network (DAN) was associated with encoding and retrieval tasks. Connectivity within DAN and between DAN and frontoparietal networks was associated with improving performance and decreasing age. In both sexes, researchers found positive correlations in retrieval accuracy and negative correlations in age to connectivity within the DAN and between DAN and frontoparietal and visual networks. Only females displayed increased between-networks at both levels of task difficulty. In males, DAN-default mode connectivity associated positively with age and negatively with performance [55]. These results underline sex-specific neural mechanisms to memory tasks that are altered with age.

A study explored sex differences in the FC basis of performance of intelligence quotient tests in young adults (17–24 years). Prediction of performance based on FC was consistently higher for females. High intelligence scores associated with FC in the superior frontal gyrus, precuneus, frontal gyrus, parahippocampal gyrus, and inferior and superior parietal lobule. In males, FC in temporal gyrus, basal ganglia, amygdala, thalamus, and precuneus were the most closely associated with high intelligence scores [56]. These results reflect different patterns of brain activity for intelligence between males and females.

Sex differences in language tasks can produce interesting evidence for divergent language processes in men and women. In a semantic decision task, females performed with a faster reaction time overall. Activation was observed equally in both sexes in the following regions: left inferior frontal gyri (IFG), left superior/middle temporal gyrus (STG), left superior parietal lobe (SPL), left inferior and middle occipital gyri (IOG), and the fusiform gyrus (FG). Dynamic interactions differed, however. Females had higher interactions IFG to STG; males were higher SPL to STG. Females exhibit more small-worldness, while males show increased transitivity, clustering coefficient, and path length. Males display higher functional segregation, while females display more functional integration [57]. While regions of activation were the same among sexes, these stark differences in clustering patterns and dynamic interactions provide evidence for different neural approaches to language tasks. These results also highlight the importance of examining multiple modes of neural activation, as each provide a different assay of information.

One team applied NISD research tactics to FC during emotion regulation. Participants were to suppress negative emotions while observing alarming photos. Males and females were equal in self-reported suppression success, but females reported more negative affect. Suppression success was associated with increased connectivity in CON in females, and in the ventral attention network for males. Based on these results, the authors conclude females use top-down regulation because they are more emotionally reactive, and males switch their focus from the negative stimulus with the ventral attention network [58]. This implies that females generate more emotion than males for the same stimulus. However, these data could also be explained by differences in emotion processing. Top-down reasoning could enable females to process and explain their emotions more readily, while also experiencing the emotion itself more deeply. The ventral attention network in males could create a distraction from the emotion, rather than the stimulus itself, so that the emotion is suppressed but not processed. These potential conclusions could also inform the different types of emotional disorders seen more commonly in females versus males.

Task-based NISD can elucidate functional effects of neuropsychiatric disease. A team of researchers conducted fMRI scans while they administered the Montreal Imaging Stress Task to unmedicated MDD patients and healthy controls. Their regions of interest include known hotspots for MDD—the amygdala, hippocampus, medial orbitofrontal cortex (OFC), NAc, and dorsolateral prefrontal cortex (dlPFC). MDD males exhibited increased connectivity to healthy controls in the dlPFC and the right frontoparietal network. Decreases in the amygdala, NAc, hippocampus, and amygdala-NAc-ACC network were observed in MDD females compared to healthy controls. These regions did not increase deactivation to stress over time in MDD females [59]. These results show important sex-case interactions and sex differences in neural stress response for individuals with MDD.

Reviews report a trend away from exploring sex differences in task-based FC research. Some researchers fail to publish data that contain sexist undertones, while others avoid developing a project on sex differences due to fear of sexism accusations or rejection from journals. Researchers may also fear their data will be used to fuel sexist propaganda [1]. Task-based functional NISD may garner negative attention if publications focus too heavily on task performance. The connectivity mechanisms for task approach provide more consistent results, which is evident in the mental rotation examples of this section, and these data provide more information on sex-specific neural processes. Researchers could also focus task-based research to specific populations, such as neuropsychiatric or neurodegenerative disease. All in all, task-based FC NISD can provide important evidence to increase understanding of neural processes and improve well-informed patient care.

4 Endocrinology

Hormones are highly linked to neurodevelopment and behavior, and hormone differences source many of the neurodivergence seen between men and women. Thus, examining how hormones and neural processes interact can increase understanding of how men and women handle various experiences of the world. The following is a survey of recent NISD that combines neuroimaging and endocrinology throughout stages of life.

Functional differences between females and males have been observed from childhood to adulthood. A study investigated whether these differences can be detected in utero by conducting resting state fMRI scans on 95 fetuses (19–40 weeks). This project developed a support vector machine learning model that classified sex with 73% accuracy. Reliable consistency was observed in that females display greater cross-region connectivity, while males exceed in connectivity within anatomically bound regions [60]. These results echo those of fMRI studies in adults, and the paper provides evidence for sex-specific neurodevelopment at very early stages in life.

A study investigated sex differences in avoidant/restrictive food intake disorder (ARFID), an eating disorder involving aversion to food, which occurs most often in young males. This project analyzed hormones related to eating, peptide YY (anti-appetite) and ghrelin (pro-appetite), as well as BOLD activation via fMRI in hunger circuitry. Interestingly, the researchers found no sex difference in hormone expression or BOLD data among 75 individuals with ARFID [61]. Despite the lack of statistical significance in its data, this paper provides examples of methods to examine appetite and feeding behavior.

A longitudinal study tracked the effect of multiple types of ELA on neurobiology in males and females. Unpredictability and threat associate with earlier menarche, deprivation associates with steeper increases in BMI, and internalizing and externalizing symptoms associate with faster pubertal tempo in females. In males, ELA was not associated with measures of puberty, physical development, or with internalizing/externalizing symptoms [62]. This research aligns with previously mentioned evidence of higher resilience in males.

The beginning stages of the NIMH Intramural Longitudinal Study of the Endocrine and Neurobiological Events Accompanying Puberty track interaction of puberty, hypothalamic-pituitary–gonadal (HPG) axis, steroid hormones, and neurodevelopment. Endocrine, gonadal, adrenal, growth, brain structure and FC are measured in children starting at eight years old. Preliminary data report functional activation in response to tasks, but further data has yet to be published [63]. This project could provide new waves of information about sex differences in adolescence and neurodevelopment.

A study investigating the effects of pregnancy on brain structure in first-time mothers, fathers, and non-parental controls located significant structural differences in only and all pregnant women after delivery. GMV reductions were observed mainly in the anterior and posterior midline, the bilateral lateral prefrontal cortex, and the bilateral temporal cortex. Interestingly, the affected regions align with areas that respond to babies postpartum, and GMV reductions lasted for at least two years postpartum [64]. These findings suggest that sustained and functional changes to female neuroanatomy as a result of motherhood.

Prolactinomas, tumors of the pituitary gland, result in exaggerated endogenous hormone fluctuations. A study employed rs-fMRI to explore whole brain FC and hormone levels in prolactinoma patients. Prolactinomas correlated with increased thalamocortical and cerebellar FC in both sexes. In females, elevated prolactin correlated with higher FC in the left thalamus and left lingual gyrus. In males, testosterone levels had a positive association with FC in the right cerebellum and left temporal fusiform cortex. Additionally, luteinizing hormone levels in males correlated with FC in the left supplementary motor area and right lingual gyrus [65]. These sex-specific hormone-connectivity interactions could explain divergent prolactinoma phenotypes between men and women.

The amygdala contains receptors for androgens, such as testosterone, and it is tied to several psychiatric conditions, including PTSD, anxiety disorders, and aggression disorders. One study investigated rs-FC in the amygdala in association with testosterone levels. In both men and women, elevated testosterone was positively associated with rs-FC between the right amygdala and right occipital gyrus, and negatively associated with agreeableness scores. In women, testosterone levels negatively associated with rs-FC between the right superior frontal gyrus and the right amygdala [66]. These results provide evidence for testosterone influences on aggression and connectivity in the amygdala.

An fMRI study expanded their results to sex-hormone-related genes and pathways. First, the study identified high complexity in the frontoparietal control and CON, and low complexity in cerebellar and sensory networks for both sexes. Females displayed greater signal complexity in all networks, especially DMN. Next, the study investigated genes related to sex differences in network complexity. The significant genes were expressed differently in cortex, estrogen-signaling pathway, biological function of neuroplasticity. The G-protein-coupled estrogen receptor 1 gene in the estrogen signaling pathway was expressed more in brain regions with higher sex differences in complexity, identifying biomarkers for the increased connectivity observed in females [31].

AD, which is more prevalent in women, may be affected by menopause-related hormone fluctuations. A study used MRI and PET to assess AD biomarkers in symptomatic perimenopausal (SM), asymptomatic perimenopausal (AP), postmenopausal (PM) women, and age-matched men (M). The results for SM and PM reported significantly higher incidence of hypometabolism, increased beta-amyloid deposition, and reduced volume in AD-susceptible brain areas, including the posterior cingulate/

precuneus, frontal, temporal, and parietal regions, compared to AP and M [67]. These findings implicate menopause-related hormonal fluctuations in the vulnerability of women to AD.

5 Conclusions and Potential Clinical Implications

A key clinical implication of investigating sex and gender differences in neuroimaging research is the potential for tailored treatment approaches for men and women. This chapter emphasizes the critical role of sex considerations in advancing clinical practice by integrating structural and functional neuroimaging. Evidence additionally highlights how sex hormones influence brain function and disease progression in various neurological conditions. These multilayered methods promise deeper insights into sex-specific neural processes and implications for diverse populations. Furthermore, findings from NISD could influence the effectiveness of medical treatment, in neuropsychiatric, developmental, and degenerative diseases. Moving forward, expanding sex-based research in neuroimaging will be essential to developing targeted, individualized treatments that optimize health outcomes for both men and women.

Acknowledgements Parts of this chapter were presented at BrainLink 2023 in PyeongChang by Professor H.W. Lee and Professor R. T. Constable. The authors would like to express special thanks to Dr. Na-Young Kim, and Dr. Heisook Lee, who organized the BrainLink 2023. This study was partly supported by the National Strategic R&D Project (RS-2023-00265524) and BK21 Plus Programs through the National Research Foundation of Korea, and the Institute of Information & communications Technology Planning & Evaluation (IITP) grant (No. RS-2022-00155966) by the Korean government (MSIT) and the Artificial Intelligence Convergence Innovation Human Resources Development programs of Ewha Womans University (H.W.Lee).

References

1. Fine C. Is there neurosexism in functional neuroimaging investigations of sex differences? Neuroethics. 2013;6:369–409.
2. Joynes CM, Bilgel M, An Y, Moghekar AR, Ashton NJ, Kac PR, et al. Sex differences in the trajectories of plasma biomarkers, brain atrophy, and cognitive decline relative to amyloid onset. Alzheimers Dement. 2025;21: e14405.
3. Chowen JA, Garcia-Segura LM. Role of glial cells in the generation of sex differences in neurodegenerative diseases and brain aging. Mech Ageing Dev. 2021;196: 111473.
4. Liu S, Seidlitz J, Blumenthal JD, Clasen LS, Raznahan A. Integrative structural, functional, and transcriptomic analyses of sex-biased brain organization in humans. Proc Natl Acad Sci U S A. 2020;117:18788–98.
5. Shafee R, Moraczewski D, Liu S, Mallard T, Thomas A, Raznahan A. A sex-stratified analysis of the genetic architecture of human brain anatomy. Nat Commun. 2024;15:8041.
6. van Eijk L, Hansell NK, Strike LT, Couvy-Duchesne B, de Zubicaray GI, Thompson PM, et al. Region-specific sex differences in the hippocampus. Neuroimage. 2020;215: 116781.

7. Brennan D, Wu T, Fan J. Morphometrical brain markers of sex difference. Cereb Cortex. 2021;31:3641–9.
8. Ingalhalikar M, Smith A, Parker D, Satterthwaite TD, Elliott MA, Ruparel K, et al. Sex differences in the structural connectome of the human brain. Proc Natl Acad Sci U S A. 2014;111:823–8.
9. Rice LC, Rochowiak RN, Plotkin MR, Rosch KS, Mostofsky SH, Crocetti D. Sex differences and behavioral associations with typically developing pediatric regional cerebellar gray matter volume. Cerebellum. 2024;23:589–600.
10. Goldstein JM, Seidman LJ, O'Brien LM, Horton NJ, Kennedy DN, Makris N, et al. Impact of normal sexual dimorphisms on sex differences in structural brain abnormalities in schizophrenia assessed by magnetic resonance imaging. Arch Gen Psychiatry. 2002;59:154–64.
11. Mitchell RH, Metcalfe AW, Islam AH, Toma S, Patel R, Fiksenbaum L, et al. Sex differences in brain structure among adolescents with bipolar disorder. Bipolar Disord. 2018;20:448–58.
12. Hu X, Zhang L, Liang K, Cao L, Liu J, Li H, et al. Sex-specific alterations of cortical morphometry in treatment-naïve patients with major depressive disorder. Neuropsychopharmacology. 2022;47:2002–9.
13. Matte Bon G, Kraft D, Comasco E, Derntl B, Kaufmann T. Modeling brain sex in the limbic system as phenotype for female-prevalent mental disorders. Biol Sex Differ. 2024;15:42.
14. Grace S, Rossetti MG, Allen N, Batalla A, Bellani M, Brambilla P, et al. Sex differences in the neuroanatomy of alcohol dependence: hippocampus and amygdala subregions in a sample of 966 people from the ENIGMA Addiction Working Group. Transl Psychiatry. 2021;11:156.
15. Perez Diaz M, Pochon JB, Ghahremani DG, Dean AC, Faulkner P, Petersen N, et al. Sex differences in the association of cigarette craving with insula structure. Int J Neuropsychopharmacol. 2021;24:624–33.
16. Tremblay C, Abbasi N, Zeighami Y, Yau Y, Dadar M, Rahayel S, et al. Sex effects on brain structure in de novo Parkinson's disease: a multimodal neuroimaging study. Brain. 2020;143:3052–66.
17. Yang Y, Hu L, Chen Y, Gu W, Xie Y, Nie S. Sex-specific imaging biomarkers for Parkinson's disease diagnosis: a machine learning analysis. J Imaging Inform Med. 2025;38:1062–75.
18. Chen J, Ragab AAY, Doyle MF, Alosco ML, Fang Y, Mez J, et al. Inflammatory protein associations with brain MRI measures: Framingham Offspring Cohort. Alzheimers Dement. 2024;20:7465–78.
19. Bath KG. Synthesizing views to understand sex differences in response to early life adversity. Trends Neurosci. 2020;43:300–10.
20. Christensen R, Chau V, Synnes A, Guo T, Ufkes S, Grunau RE, et al. Preterm sex differences in neurodevelopment and brain development from early life to 8 years of age. J Pediatr. 2025;276: 114271.
21. McManus E, Haroon H, Duncan NW, Elliott R, Muhlert N. Hippocampal and limbic microstructure changes associated with stress across the lifespan: a UK biobank study. Sci Rep. 2024;14:21735.
22. Cornwell H, Toschi N, Hamilton-Giachritsis C, Staginnus M, Smaragdi A, Gonzalez-Madruga K, et al. Identifying structural brain markers of resilience to adversity in young people using voxel-based morphometry. Dev Psychopathol. 2023;35:2302–14.
23. Pan N, Yang C, Suo X, Shekara A, Hu S, Gong Q, et al. Sex differences in the relationship between brain gray matter volume and psychological resilience in late adolescence. Eur Child Adolesc Psychiatry. 2024;33:1057–66.
24. Ancelin ML, Carrière I, Artero S, Maller JJ, Meslin C, Dupuy AM, et al. Structural brain alterations in older adults exposed to early-life adversity. Psychoneuroendocrinology. 2021;129: 105272.
25. Murphy DG, DeCarli C, McIntosh AR, Daly E, Mentis MJ, Pietrini P, et al. Sex differences in human brain morphometry and metabolism: an in vivo quantitative magnetic resonance imaging and positron emission tomography study on the effect of aging. Arch Gen Psychiatry. 1996;53:585–94.

26. Kim HJ, Kim REY, Kim S, Kim SA, Kim SE, Lee SK, et al. Sex differences in deterioration of sleep properties associated with aging: a 12-year longitudinal cohort study. J Clin Sleep Med. 2021;17:964–72.
27. Kim HJ, Kim REY, Kim S, Lee SK, Lee HW, Shin C. Earlier chronotype in midlife as a predictor of accelerated brain aging: a population-based longitudinal cohort study. Sleep. 2023;46:zsad108.
28. Zhang X, Liang M, Qin W, Wan B, Yu C, Ming D. Gender differences are encoded differently in the structure and function of the human brain revealed by multimodal MRI. Front Hum Neurosci. 2020;14:244.
29. Spreng RN, Dimas E, Mwilambwe-Tshilobo L, Dagher A, Koellinger P, Nave G, et al. The default network of the human brain is associated with perceived social isolation. Nat Commun. 2020;11:6393.
30. Serio B, Hettwer MD, Wiersch L, Bignardi G, Sacher J, Weis S, et al. Sex differences in intrinsic functional cortical organization reflect differences in network topology rather than cortical morphometry. bioRxiv. 2023.
31. Zhao CL, Hou W, Jia Y, Sahakian BJ, Luo Q. Sex differences of signal complexity at resting-state functional magnetic resonance imaging and their associations with the estrogen-signaling pathway in the brain. Cogn Neurodyn. 2024;18:973–86.
32. Yao YJ, Chen Y, Li CR. Hypothalamic connectivities and self-evaluated aggression in young adults. bioRxiv. 2024.
33. Williamson J, James SA, Mukli P, Yabluchanskiy A, Wu DH, Sonntag W, et al. Sex difference in brain functional connectivity of hippocampus in Alzheimer's disease. Geroscience. 2024;46:563–72.
34. Williamson J, Yabluchanskiy A, Mukli P, Wu DH, Sonntag W, Ciro C, et al. Sex differences in brain functional connectivity of hippocampus in mild cognitive impairment. Front Aging Neurosci. 2022;14: 959394.
35. Wheelock MD, Hect JL, Hernandez-Andrade E, Hassan SS, Romero R, Eggebrecht AT, et al. Sex differences in functional connectivity during fetal brain development. Dev Cogn Neurosci. 2019;36: 100632.
36. Eyre M, Fitzgibbon SP, Ciarrusta J, Cordero-Grande L, Price AN, Poppe T, et al. The developing human connectome project: typical and disrupted perinatal functional connectivity. Brain. 2021;144:2199–213.
37. Fan F, Liao X, Lei T, Zhao T, Xia M, Men W, et al. Development of the default-mode network during childhood and adolescence: a longitudinal resting-state fMRI study. Neuroimage. 2021;226: 117581.
38. Gao L, Cao Y, Zhang Y, Liu J, Zhang T, Zhou R, et al. Sex differences in the flexibility of dynamic network reconfiguration of autism spectrum disorder based on multilayer network. Brain Imaging Behav. 2024;18:1172–85.
39. Tavares V, Fernandes LA, Antunes M, Ferreira H, Prata D. Sex fifferences in functional connectivity between resting state brain networks in autism spectrum disorder. J Autism Dev Disord. 2022;52:3088–101.
40. Lawrence KE, Hernandez LM, Bowman HC, Padgaonkar NT, Fuster E, Jack A, et al. Sex differences in functional connectivity of the salience, default mode, and central executive networks in youth with ASD. Cereb Cortex. 2020;30:5107–20.
41. Padgaonkar NT, Lawrence KE, Hernandez LM, Green SA, Galván A, Dapretto M. Sex differences in internalizing symptoms and amygdala functional connectivity in neurotypical youth. Dev Cogn Neurosci. 2020;44: 100797.
42. Borst B, Jovanovic T, House SL, Bruce SE, Harnett NG, Roeckner AR, et al. Sex differences in response inhibition-related neural predictors of posttraumatic stress disorder in civilians with recent trauma. Biol Psychiatry Cogn Neurosci Neuroimaging. 2024;9:668–80.
43. Mitchell RHB, Grigorian A, Robertson A, Toma S, Luciw NJ, Karthikeyan S, et al. Sex differences in cerebral blood flow among adolescents with bipolar disorder. Bipolar Disord. 2024;26:33–43.

44. Ravichandran S, Bhatt RR, Pandit B, Osadchiy V, Alaverdyan A, Vora P, et al. Alterations in reward network functional connectivity are associated with increased food addiction in obese individuals. Sci Rep. 2021;11:3386.
45. Koenig KA, Lowe MJ, Lin J, Sakaie KE, Stone L, Bermel RA, et al. Sex differences in resting-state functional connectivity in multiple sclerosis. AJNR Am J Neuroradiol. 2013;34:2304–11.
46. Howard EF. Exploring the relationship between structural myelin mapping and cognition in individuals with a history of traumatic brain injury. The Ohio State University; 2024.
47. Licata AE, Zhao Y, Herrmann O, Hillis AE, Desmond J, Onyike C, et al. Sex differences in effects of tDCS and language treatments on brain functional connectivity in primary progressive aphasia. Neuroimage Clin. 2023;37: 103329.
48. Shanmugan S, Seidlitz J, Cui Z, Adebimpe A, Bassett DS, Bertolero MA, et al. Sex differences in the functional topography of association networks in youth. Proc Natl Acad Sci U S A. 2022;119: e2110416119.
49. Weis S, Patil KR, Hoffstaedter F, Nostro A, Yeo BTT, Eickhoff SB. Sex classification by resting state brain connectivity. Cereb Cortex. 2020;30:824–35.
50. Zhang K, Fang H, Li Z, Ren T, Li BM, Wang C. Sex differences in large-scale brain network connectivity for mental rotation performance. Neuroimage. 2024;298: 120807.
51. Weiss E, Siedentopf CM, Hofer A, Deisenhammer EA, Hoptman MJ, Kremser C, et al. Sex differences in brain activation pattern during a visuospatial cognitive task: a functional magnetic resonance imaging study in healthy volunteers. Neurosci Lett. 2003;344:169–72.
52. Thomsen T, Hugdahl K, Ersland L, Barndon R, Lundervold A, Smievoll AI, et al. Functional magnetic resonance imaging (fMRI) study of sex differences in a mental rotation task. Med Sci Monit. 2000;6:1186–96.
53. Satterthwaite TD, Wolf DH, Roalf DR, Ruparel K, Erus G, Vandekar S, et al. Linked sex differences in cognition and functional connectivity in youth. Cereb Cortex. 2015;25:2383–94.
54. Esiaka DK, Strothkamp S, Banafsheh A, Broster L, Powell DK, Jicha G, et al. Differential speed and accuracy trade-off in working memory retrieval and bilateral precuneus between elder men and women. bioRxiv. 2024:2024.10.05.616832.
55. Subramaniapillai S, Rajagopal S, Ankudowich E, Pasvanis S, Misic B, Rajah MN. Age- and episodic memory-related differences in task-based functional connectivity in women and men. J Cogn Neurosci. 2022;34:1500–20.
56. Jiang R, Calhoun VD, Fan L, Zuo N, Jung R, Qi S, et al. Gender differences in connectome-based predictions of individualized intelligence quotient and sub-domain scores. Cereb Cortex. 2020;30:888–900.
57. Xu M, Liang X, Ou J, Li H, Luo YJ, Tan LH. Sex differences in functional brain networks for language. Cereb Cortex. 2020;30:1528–37.
58. Stoica T, Knight LK, Naaz F, Patton SC, Depue BE. Gender differences in functional connectivity during emotion regulation. Neuropsychologia. 2021;156: 107829.
59. Dong D, Ironside M, Belleau EL, Sun X, Cheng C, Xiong G, et al. Sex-specific neural responses to acute psychosocial stress in depression. Transl Psychiatry. 2022;12:2.
60. Cook KM, De Asis-Cruz J, Lopez C, Quistorff J, Kapse K, Andersen N, et al. Robust sex differences in functional brain connectivity are present in utero. Cereb Cortex. 2023;33:2441–54.
61. Getachew ED, Van De Water AL, Kuhnle M, Hauser K, Kambanis PE, Plessow F, et al. Absence of sex differences identified in food motivation pathways in youth with avoidant/restrictive food intake disorder (ARFID). J Endocr Soc. 2021;5:A2–3.
62. Ho TC, Buthmann J, Chahal R, Miller JG, Gotlib IH. Exploring sex differences in trajectories of pubertal development and mental health following early adversity. Psychoneuroendocrinology. 2024;161: 106944.
63. Cole KM, Wei SM, Martinez PE, Nguyen TV, Gregory MD, Kippenhan JS, et al. The NIMH intramural longitudinal study of the endocrine and neurobiological events accompanying puberty: protocol and rationale for methods and measures. Neuroimage. 2021;234: 117970.
64. Hoekzema E, Barba-Müller E, Pozzobon C, Picado M, Lucco F, García-García D, et al. Pregnancy leads to long-lasting changes in human brain structure. Nat Neurosci. 2017;20:287–96.

65. Yao S, Lin P, Vera M, Akter F, Zhang RY, Zeng A, et al. Hormone levels are related to functional compensation in prolactinomas: a resting-state fMRI study. J Neurol Sci. 2020;411: 116720.
66. Kogler L, Müller VI, Moser E, Windischberger C, Gur RC, Habel U, et al. Testosterone and the amygdala's functional connectivity in women and men. J Clin Med. 2023;12:6501.
67. Mosconi L, Berti V, Quinn C, McHugh P, Petrongolo G, Varsavsky I, et al. Sex differences in Alzheimer risk: brain imaging of endocrine vs chronologic aging. Neurology. 2017;89:1382–90.

Part IV
Environmental and Lifespan Considerations

Programming Effects of Maternal Stress on the Brain and Circadian Timing System: Focus on Sex Differences

Sooyoung Chung

Abstract Maternal stress has long-lasting influences on the brain functions of offspring, and several brain regions have been proposed to mediate such programming. Prenatal stress can have long-term consequences on cognition and emotion through impacts on several brain regions, such as the hippocampus, amygdala, ventral midbrain, and paraventricular nucleus of the hypothalamus. Exposing pregnant mice to chronic restraint stress had prolonged effects on the suprachiasmatic nucleus (SCN) of the hypothalamus, which bears the central pacemaker for mammalian circadian rhythms, of offspring. The widespread effects of circadian disruptions caused by a misprogrammed clock may have further impacts on metabolic and mental health in later life. Effect of prenatal stress on physiology and behavior have been mainly studied in male offspring in rodent models, but on the other hand, gender differences have been investigated through human studies. Therefore, in this article, we will discuss the various effects of maternal stress, differences by gender, and various studies supporting this.

Keywords Maternal stress · Programming effect · Sex differences · Circadian rhythm

1 Introduction

Stress encompasses various stimuli capable of disturbing an individual's homeostasis. In response to such stressors, individuals employ a hormonal control system known as the stress response axis. Maternal stress, occurring at a time of heightened vulnerability, particularly influences fetal development. The fetus assimilates information from the external environment before birth and undergoes a programming process to adapt during its developmental stages.

S. Chung (✉)
Department of Brain and Cognitive Sciences, Scranton College, Ewha Womans University, Seoul, Korea
e-mail: csooy@ewha.ac.kr

H. Lee et al. (eds.), *Sex, Gender, and Emerging Technology in Healthcare: Mitigating Bias and Fostering Equity*, https://doi.org/10.1007/978-981-95-2070-1_6

A suitable level of stress aids an individual in adapting to the external environment. However, prolonged exposure to stress can lead to various issues and ultimately elevate the risk of various diseases. It is widely acknowledged that maternal stress influences the stress reactivity of offspring, potentially giving rise to ongoing challenges in alignment with the external environment even after birth [1, 2]. Furthermore, studies indicate that mice exposed to stress during birth may encounter cognitive and emotional challenges persisting into adulthood. More recent research suggests similar effects on neurodevelopment in both humans and animals [3, 4]. Notably, existing studies on this topic have predominantly focused on males in mouse models, despite clear gender differences identified in human studies. Therefore, this article delves into animal experimental models and clinical studies related to maternal stress, examining gender differences and the diverse array of supporting research.

2 Stress in the Womb

The environment within the womb is a critical societal concern as all members of society originate from it. Consequently, research on the women responsible for creating these special environments becomes even more imperative. When an individual experiences stress, hormonal changes ensue through the activation of the hypothalamic-pituitary adrenal (HPA) axis. Glucocorticoids (GCs), steroid hormones secreted from the adrenal gland, play a key role in inducing various physiological responses to prepare for stressful situations.

To prepare for stressful situations, GCs play a crucial role in promoting energy production by regulating metabolism and the cardiovascular system. They accumulate energy by conserving basic resources utilized for growth, reproduction, and immune responses. Additionally, GCs are recognized for their regulatory effects on cognitive function, mood, sleep–wake cycles, and the capacity to reset or synchronize biological rhythms.

While an appropriate concentration of stress hormones is physiologically essential for preparing the body to face stress, continuous stress-induced hyperactivity of the stress axis poses a significant challenge, affecting overall physiology and behavior. Early-life stress, in particular, is also believed to have profound and lasting impacts. In essence, the process by which non-genetic factors exert their influence early in life, leading to the permanent organization or imprinting of physiological systems, is commonly referred to as perinatal 'programming' [5].

Stress experienced by a pregnant mother elevates GC levels, directly transmitting them to the fetus through the placenta. Normally, the placental enzyme 11β-hydroxysteroid dehydrogenase 2 serves as a protective barrier against excessive circulating GCs [6, 7]. However, persistently elevated GC levels can lead to fetal overexposure to GCs. The repercussions of this overexposure endure into adulthood, potentially causing issues like reduced cognitive function, diminished attention, heightened anxiety, and an increased risk of metabolic diseases [3].

3 Research on the Diverse Effects of Maternal Stress: Animal Studies

Among animal studies exploring the effects of maternal stress, the oldest investigation focuses on the activity of the stress response axis. In simpler terms, the stress experienced by a mother influences the fetus, shaping the stress response to external environmental factors after birth. Traditional studies in this field often employed simple physical stress models with rodents, particularly the rat model born under prenatal restraint stress (PNRS), being the most representative. PNRS is associated with increased anxiety-like behavior and HPA axis hyperactivity in offspring and it was also believed to induce alteration in the rhythm of circadian stress hormones [8]. Furthermore, research has demonstrated that this stress model can induce lasting modifications in the sensitivity of the HPA axis and the sympathetic nervous system to stress, consequently influencing catecholamine levels. These findings strongly suggest that prenatal stress can have a lasting impact on both short- and long-term stress reactivity in offspring [9].

A similar mouse model involves the use of ICR mice subjected to prenatal restraint stress. Mice born under stress in this model are more prone to exhibiting anxious behavior, particularly when exposed to chronic repeated stress [10]. In accordance with increased anxiety, maintenance of basal HPA activity was attenuated in this animal model. We can conclude that exposure to maternal stress in the womb resulted in differential adaptive responses to chronic stress in adult male offspring. These findings offer significant insights into the correlation between maternal stress and anxiety-related emotional disorders.

Moreover, several experiments utilizing this model have revealed that mice born under maternal stress showed cognitive problems, such as difficulties in forming hippocampus-dependent place memories and amygdala-dependent fear memories [11, 12]. Additionally, these animals exhibit hyperactivity akin to attention deficit hyperactivity disorder (ADHD), coupled with the hyperactivation and alteration of the midbrain dopamine system [13].

Another frequently employed early life stress model is maternal separation, which is known to elevate anxiety and induce depression-like behavior in adult offspring. However, the interpretation of outcomes is complicated by inconsistent results, with reports indicating both instances of observable effects on behavior and cases where no discernible impact is observed [14]. The complexity also arises from the inherent difficulty of this animal model in faithfully replicating the neglect and abuse experienced by human offspring. Notably, research suggests that maternal care increases post-separation, thereby potentially mitigating the direct impact of the separation itself [15].

When investigating maternal stress through animal models, assessing similarity becomes challenging even when employing stress-inducing factors with characteristics akin to those experienced by humans. The different cognitive processes in the rodent brain compared with human brain add complexity and confusion to the result. Instead of comparing the outcomes of various forms of maternal stress in animals, many researchers have adopted approaches focused on increasing fetal GC stimulation without considering the intricate nature of these stressors. This is commonly achieved by administering dexamethasone, a synthetic GC, during pregnancy in animal models. The widespread use of GC administration approach in experiments reflects its efficacy in bypassing the complexities associated with diverse stressors.

Administration of the synthetic glucocorticoid dexamethasone during pregnancy has been associated with long-term adverse effects, such as weight loss, alterations in neuroendocrine and autonomic stress responses, stress-related behavioral disorders, and neuronal cell death [16]. Fetal exposure to DEX induces increased expression of hippocampal *CRH* and *CRHR1* genes through DNA hypomethylation, which can lead to depression-like behavior [17]. These results may provide evidence that excessive exposure to glucocorticoids, a stress hormone, during development is a potential cause of various negative outcomes associated with maternal stress. However, these studies have a limitation: unlike the immune system activation observed in natural stress responses, treatment with synthetic glucocorticoids suppresses various aspects of the maternal immune system.

Recent studies suggest that maternal immune activation is also a significant mediator of stress. Therefore, researchers have conducted experimental studies comparing models that promote maternal inflammatory responses with models involving fetal exposure to DEX [18].

4 Effect of Maternal Stress, Animal Studies to Understand Sexual Difference

Most studies investigating maternal stress have primarily involved male animals. However, recognizing the importance of assessing potential transgenerational effects based on the nature of maternal stress, there's a recent trend towards employing female models in experiments. Nevertheless, research involving both male and female offspring remains relatively scarce.

Maternal nutritional stress induced by a high-fat diet or calorie restriction has been linked to behavioral changes resembling ADHD, autism spectrum disorder (ASD), anxiety, and depression. A study conducted in rodents investigated the impact of a high-calorie diet during pregnancy on both male and female offspring, revealing heightened anxiety levels in both genders [18].

Exposure to social stress triggers depression-like behavior in pregnant mothers. To delve into these effects and explore gender disparities, researchers developed a mouse model. They observed heightened anxiety-like behavior in both male and female offspring of stressed mothers, with male offspring additionally exhibiting cognitive impairments in spatial recognition and temporal order memory [19]. These results parallel previous animal studies and research findings on cognitive developmental changes specifically observed in boys. Furthermore, a study employed an animal model where pregnant mice were exposed to simultaneous environmental pollutants and stress, resulting in significant activation of the maternal immune system. Notably, this combined exposure adversely impacts social behavior and brain circuit connectivity specifically in male offspring, implicating microglial cell dysfunction in this process [20]. As previously mentioned, maternal immune activation is also recognized as a key mediator in transmitting maternal stress to offspring. In rodents, this activation initiates several pathways within the fetal brain, such as mRNA translation, ribosome biogenesis, and stress signaling. Notably, gender differences have been observed in these processes [21].

The research findings indicating sex-dependent effects suggest that males and females may respond differently to identical maternal stressors. This underscores the necessity of accounting for gender differences in animal experiments, as these variations could significantly influence the outcomes of the studies.

5 Effects of Maternal Stress, Clinical Studies and Gender Differences

In human research, stress types are inherently diverse, with social and psychological factors playing crucial roles. Unlike animal studies, clinical research on maternal stress includes both men and women, necessitating investigation into gender differences. However, not all studies yield consistent results in this regard, making it challenging to obtain conclusive findings on gender difference.

One of the well-documented effects of intrauterine stress on gender is the differential vulnerability of male and female fetuses to physiological stress. Male fetuses are notably more susceptible to intrauterine loss during period of severe stress, resulting in lower primary survival rates compared to females. In contrast, the female placenta exhibits greater resilience and responsiveness to stress signals, such as maternal glucocorticoid concentrations, resulting in diverse growth patterns. Male fetuses often have fewer resources allocated for growth and may exhibit reduced responses to adversity. Historical research has shown that under extreme stress conditions, such as famine, male birth ratios tend to decrease, and recovery of the sex ratio is only observed once these stressors abate [22].

Maternal stress has been associated with distinct outcomes in children based on their gender. Boys exposed to maternal stress tend to exhibit increased vulnerability to conditions such as asthma and hyperactivity [23], whereas girls may be more prone to experiencing emotional symptoms like depression and anxiety [24].

Gender differences in the development of neurodevelopmental problems and neurodegenerative diseases, as emphasized by Hanamsagar and Bilbo [25], highlight the importance of prenatal environment in shaping distinct outcomes for men and women. However, the underlying causes of these differences remain unclear.

6 Multiple Response Systems that Can Mediate the Effects of Maternal Stress and the Basis for Gender Differences

The significance of stress biomarkers such as cortisol and pro-inflammatory cytokines in transmitting maternal stress to offspring is well recognized. Cortisol, a pivotal stress hormone, exerts diverse effects with profound impacts on the endocrine, immune, and nervous systems. Notably, HPA axis, a neuroendocrine network regulating stress hormone secretion, is modulated by gonadal hormones like testosterone and estradiol, contributing to gender differences in the human HPA axis. Typically, females demonstrate heightened reactivity in the stress axis, coupled with delayed termination of stress responses and weak feedback inhibition of the HPA axis. These findings suggest that biological sex also plays a role in determining the magnitude of the stress response itself [26].

Additionally, maternal stress is thought to increase cytokine production, promoting inflammatory conditions in the placenta, thereby changing HPA activity in offspring. Glucocorticoids, which are stress hormones, directly regulate immune responses, and it is already well known that the immune responses of men and women show gender differences, so these differences in immune response susceptibility may also be a mechanism that mediates gender differences [27].

Recent studies have revealed that maternal stress can alter the microbial community of offspring by influencing the microbial composition in the mother's intestinal and vaginal compartments. Interestingly, this process leads to distinct microbial communities in male and female offspring. These findings suggest that the intestinal microbial community also serves as a conduit through which maternal stress experiences are transmitted, while also indicating the possible emergence of gender differences through this mechanism [28].

7 Maternal Stress and Circadian Rhythm

The circadian rhythm, with a period of approximately 24 h, orchestrates biological oscillations that synchronize behavioral and physiological processes, facilitating adaptation to daily environmental changes. This rhythm is governed by an internal timing system reliant on genetic components known as the circadian molecular clock, and its manifestations can be assessed through gene expression patterns [29].

The association between the stress system and the circadian system in adulthood is well established. Adrenal glucocorticoids, pivotal stress hormones, not only act as key mediators of the stress response but also exhibit rhythmic secretion patterns throughout the day, independent of stress levels. This rhythmic pattern has been shown to be regulated by the biological clock [30]. Particularly noteworthy is the role of rhythmic glucocorticoid signaling in synchronizing and stabilizing locomotor activity and circadian rhythms across various peripheral tissues.

A recent study offers direct evidence supporting the idea that maternal stress during pregnancy profoundly affects the development of the SCN in the hypothalamus, consequently exerting long-term effects on the offspring's circadian system [31]. Given the co-occurrence of other brain diseases, sleep disorders, and circadian cycle disruptions resulting from prenatal stress, it is plausible that the circadian system mediates various adverse effects of maternal stress on the physiological and mental health of offspring. This suggests a potential avenue for further investigation into the underlying mechanisms.

Recent studies have also highlighted potential gender disparities in the operation of the biological clock [32]. Moreover, owing to the interplay between sex hormones and biological rhythms, discrepancies in gender may manifest in the amplitude of these rhythms or in the ability to adapt to external stimuli. Consequently, there is a possibility that gender variations in the impacts of maternal stress are mediated through the circadian rhythm system.

8 Problems in Research on Gender Differences

Most animal studies involving maternal stress predominantly utilize male animals due to the complexities associated with the reproductive cycle in females. Emotional disorders, memory formation, and changes in the stress axis—all of which are stress phenotypes—depend on the timing of the reproductive cycle in female animals. Rodents, frequently used in experiments, have short and fluctuating reproductive cycles (typically 3–4 days). Therefore, efforts are made to synchronize the female cycle or conduct experiments exclusively on females at specific stages, necessitating additional resources to achieve comparable results.

Another challenge arises from the fact that behavioral tests commonly used to measure various changes have primarily been developed for male subjects, raising concerns about their accuracy in assessing changes in females. There is a pressing need to rectify the underrepresentation of female in research by developing new, sensitive behavioral analysis methods that are applicable to both genders.

Furthermore, there is an ongoing debate about the comparability of human and animal studies conducted thus far. It is highly likely that the effects of maternal stress vary depending on the type and process of occurrence of stressors. Human research reflects a wide range of stressors experienced by pregnant women, including significant mental stress. Gender differences observed in humans primarily consider effects on a single fetus. However, it remains uncertain whether the gender differences observed in rodents, which typically have multiple fetuses, share the same underlying mechanisms. Moreover, it is unclear whether stress types experienced by humans can exert similar effects in animals due to differences in cognitive processes.

Ultimately, the primary issue lies in the dearth of foundational research supporting investigations into gender differences in response to maternal stress in humans.

There is an urgent need for systematic research that incorporates key factors mediating maternal stress to better understand these differences and inform targeted interventions.

Acknowledgements This study was supported by the Ewha Womans University Research Grant.

References

1. Harris A, Seckl J. Glucocorticoids, prenatal stress and the programming of disease. Horm Behav. 2011;59:279–89.
2. Maccari S, Polese D, Reynaert ML, Amici T, Morley-Fletcher S, Fagioli F. Early-life experiences and the development of adult diseases with a focus on mental illness: the human birth theory. Neuroscience. 2017;342:232–51.
3. Goldstein JM, Hale T, Foster SL, Tobet SA, Handa RJ. Sex differences in major depression and comorbidity of cardiometabolic disorders: impact of prenatal stress and immune exposures. Neuropsychopharmacology. 2019;44:59–70.
4. Sheng JA, Tan SML, Hale TM, Handa RJ. Androgens and their role in regulating sex differences in the hypothalamic/pituitary/adrenal axis stress response and stress-related behaviors. Androg Clin Res Ther. 2021;2:261–74.
5. Barker DJ. The fetal and infant origins of adult disease. BMJ. 1990;301:1111.
6. Welberg LA, Seckl JR, Holmes MC. Inhibition of 11beta-hydroxysteroid dehydrogenase, the foeto-placental barrier to maternal glucocorticoids, permanently programs amygdala GR mRNA expression and anxiety-like behaviour in the offspring. Eur J Neurosci. 2000;12:1047–54.
7. Seckl JR, Holmes MC. Mechanisms of disease: glucocorticoids, their placental metabolism and fetal 'programming' of adult pathophysiology. Nat Clin Pract Endocrinol Metab. 2007;3:479–88.
8. Darnaudéry M, Maccari S. Epigenetic programming of the stress response in male and female rats by prenatal restraint stress. Brain Res Rev. 2008;57:571–85.
9. Maccari S, Darnaudery M, Morley-Fletcher S, Zuena AR, Cinque C, Van Reeth O. Prenatal stress and long-term consequences: implications of glucocorticoid hormones. Neurosci Biobehav Rev. 2003;7:119–27.
10. Chung S, Son GH, Park SH, Park E, Lee KH, Geum D, et al. Differential adaptive responses to chronic stress of maternally stressed male mice offspring. Endocrinology. 2005;146:3202–10.
11. Son GH, Geum D, Chung S, Kim EJ, Jo JH, Kim CM, et al. Maternal stress produces learning deficits associated with impairment of NMDA receptor-mediated synaptic plasticity. J Neurosci. 2006;26:3309–18.
12. Lee EJ, Son GH, Chung S, Lee S, Kim J, Choi S, et al. Impairment of fear memory consolidation in maternally stressed male mouse offspring: evidence for nongenomic glucocorticoid action on the amygdala. J Neurosci. 2011;31:7131–40.
13. Son GH, Chung S, Geum D, Kang SS, Choi WS, Kim K, et al. Hyperactivity and alteration of the midbrain dopaminergic system in maternally stressed male mice offspring. Biochem Biophys Res Commun. 2007;352:823–9.
14. Murthy S, Gould E. Early life stress in rodents: animal models of illness or resilience? Front Behav Neurosci. 2018;12:157.
15. Millstein RA, Holmes A. Effects of repeated maternal separation on anxiety- and depression-related phenotypes in different mouse strains. Neurosci Biobehav Rev. 2007;31:3–17.
16. Wang J, Chen F, Zhu S, Li X, Shi W, Dai Z, et al. Adverse effects of prenatal dexamethasone exposure on fetal development. J Reprod Immunol. 2022;151: 103619.

17. Xu YJ, Sheng H, Wu TW, Bao QY, Zheng Y, Zhang YM, et al. CRH/CRHR1 mediates prenatal synthetic glucocorticoid programming of depression-like behavior across 2 generations. FASEB J. 2018;32:4258–69.
18. Sheng JA, Handa RJ, Tobet SA. Evaluating different models of maternal stress on stress-responsive systems in prepubertal mice. Front Neurosci. 2023;17:1292642.
19. Scarborough J, Mueller FS, Weber-Stadlbauer U, Mattei D, Opitz L, Cattaneo A, et al. A novel murine model to study the impact of maternal depression and antidepressant treatment on biobehavioral functions in the offspring. Mol Psychiatry. 2021;26:6756–72.
20. Block CL, Eroglu O, Mague SD, Smith CJ, Ceasrine AM, Sriworarat C, et al. Prenatal environmental stressors impair postnatal microglia function and adult behavior in males. Cell Rep. 2022;40: 111161.
21. Kalish BT, Kim E, Finander B, Duffy EE, Kim H, Gilman CK, et al. Maternal immune activation in mice disrupts proteostasis in the fetal brain. Nat Neurosci. 2021;24:204–13.
22. Sandman CA, Glynn LM, Davis EP. Is there a viability-vulnerability tradeoff? Sex differences in fetal programming. J Psychosom Res. 2013;75:327–35.
23. Sutherland S, Brunwasser SM. Sex differences in vulnerability to prenatal stress: a review of the recent literature. Curr Psychiatry Rep. 2018;20:102.
24. Quarini C, Pearson RM, Stein A, Ramchandani PG, Lewis G, Evans J. Are female children more vulnerable to the long-term effects of maternal depression during pregnancy? J Affect Disord. 2016;189:329–35.
25. Hanamsagar R, Bilbo SD. Sex differences in neurodevelopmental and neurodegenerative disorders: focus on microglial function and neuroinflammation during development. J Steroid Biochem Mol Biol. 2016;160:127–33.
26. Heck AL, Handa RJ. Sex differences in the hypothalamic–pituitary–adrenal axis' response to stress: an important role for gonadal hormones. Neuropsychopharmacology. 2019;44:45–58.
27. Klein SL, Flanagan KL. Sex differences in immune responses. Nat Rev Immunol. 2016;16:626–38.
28. Yeramilli V, Cheddadi R, Shah J, Brawner K, Martin C. A review of the impact of maternal prenatal stress on offspring microbiota and metabolites. Metabolites. 2023;13:535.
29. Chung S, Son GH, Kim K. Circadian rhythm of adrenal glucocorticoid: its regulation and clinical implications. Biochim Biophys Acta. 2011;1812:581–91.
30. Son GH, Chung S, Choe HK, Kim HD, Baik SM, Lee H, et al. Adrenal peripheral clock controls the autonomous circadian rhythm of glucocorticoid by causing rhythmic steroid production. Proc Natl Acad Sci U S A. 2008;105:20970–5.
31. Yun S, Lee EJ, Choe HK, Son GH, Kim K, Chung S. Programming effects of maternal stress on the circadian system of adult offspring. Exp Mol Med. 2020;52:473–84.
32. Boyd HM, Frick KM, Kwapis JL. Connecting the dots: potential interactions between sex hormones and the circadian system during memory consolidation. J Biol Rhythms. 2023;38:537–55.

Sex and Gender Differences in Drug Safety

Na-Young Jeong and Nam-Kyong Choi

Abstract This chapter explores how biological sex differences influence drug safety and efficacy, underscoring the need to incorporate sex-specific considerations across all stages of drug development. Historically, the underrepresentation of females in preclinical and clinical studies has contributed to gaps in knowledge that compromise treatment effectiveness and increase adverse drug reactions in women. The chapter outlines key differences in pharmacokinetics and pharmacodynamics between males and females—such as variations in body composition, enzyme activity, and hormone levels—leading to sex-specific differences in drug absorption, metabolism, and response. It also presents evidence from pharmacovigilance data highlighting distinct patterns in adverse event reporting by sex and gender. In response, regulatory frameworks have increasingly emphasized the inclusion of sex-disaggregated data and adequate female representation in clinical trials. Methodological strategies for integrating sex and gender perspectives in both clinical research and real-world evidence generation are discussed. By addressing sex as a biological variable from the earliest stages of drug development through post-marketing surveillance, the chapter emphasizes the importance of improving drug safety, advancing personalized medicine, and achieving more equitable healthcare outcomes.

Keywords Drug safety · Sex and/or gender differences · Adverse drug reactions · Representation in clinical trials · Pharmacovigilance

---

N.-Y. Jeong
Health Science Convergence Research Institute, Ewha Womans University, Seoul, Republic of Korea

N.-K. Choi (✉)
Department of Health Convergence, College of Science and Industry Convergence, Ewha Womans University, Seoul, Republic of Korea
e-mail: nchoi@ewha.ac.kr

Graduate School of Industrial Pharmaceutical Science, College of Pharmacy, Ewha Womans University, Seoul, Republic of Korea

H. Lee et al. (eds.), *Sex, Gender, and Emerging Technology in Healthcare: Mitigating Bias and Fostering Equity*, https://doi.org/10.1007/978-981-95-2070-1_7

1 Introduction

For decades, most pharmaceutical clinical studies have predominantly involved male participants, and the majority of preclinical research has almost exclusively incorporated male cells or animals. The knowledge gained from studies focusing on male participants was generally assumed to be applicable to the entire population. This lack of female representation in drug clinical trials has created gaps in medical knowledge, leading to delayed diagnoses or misdiagnoses in female patients, and the development of drugs with lower efficacy and higher toxicity for women. Consequently, some drugs, primarily tested on males and approved based on that data, were withdrawn due to safety issues in women. According to a report published by the U.S. General Accounting Office (GAO), between January 1, 1997, and December 31, 2000, 8 out of 10 drugs withdrawn from the U.S. market by the Food and Drug Administration (FDA) had more adverse health effects on women than on men [1]. This situation occurred because sex was not considered a biological variable in preclinical trials, and clinical trials primarily targeted men. Of the eight drugs withdrawn, four were prescribed more frequently to women, while the remaining four, despite being widely prescribed to both sexes, showed higher risks in women. Major health issues raised included valvular heart disease and Torsades de Pointes.

These issues have sparked interest in the differences in drug efficacy and safety between males and females. Various evidences in medical literature present sex-based physiological and pathophysiological differences that affect the development, progression, presentation, and symptoms of diseases. Especially during drug intake, differences in metabolism and liver function between males and females affect the absorption and disposition of pharmacological agents, potentially impacting therapeutic safety and efficacy. These differences are observed in various organ systems, including the immune and cardiovascular systems. Women tend to have lower metabolic rates per body weight and higher body fat percentages, which can affect drug distribution and metabolism rates. Furthermore, changes in hormone levels due to pregnancy and menstrual cycles in women can also influence drug effects. Thus, physiological differences based on sex can lead to variations in the efficacy and side effects of the same medication. Beyond biological differences in sex and gender inequalities shaped by personal, social, environmental, psychological, and behavioral factors can disproportionately affect women's work, income, and perceived roles. These inequalities often lead to an unequal burden of caregiving responsibilities and related adverse health impacts. Sex and gender are not interchangeable but independent factors that can individually or synergistically impact health.

This chapter primarily focuses on sex differences to understand their impact on medication safety and clinical research practices. Additionally, it will explore gender-related disparities to find out how sex and gender interact to impact health outcome in various ways.

2 Sex-Specific Influence on Drug Safety

2.1 Pharmacokinetic and Pharmacodynamic Differences Based on Sex and Gender

When a patient is administered a drug, the intensity and duration of its pharmacological responses within the body are determined by pharmacokinetics (PK) -the process of absorption, distribution, metabolism, and excretion (i.e., elimination) of the drug within the body, and pharmacodynamics (PD)—the action of the drug within the body. Differences in response to drug therapy occur due to various structural and physiological differences between males and females, including differences in body weight, height, body surface area, total body water, plasma volume, organ blood flow, and extracellular and intracellular water [2–4] (Table 1). This indicates that there are sex-based differences in PK and PD (Table 2). Differences in PK between males and females can arise from various factors such as the characteristics of organs involved in absorption, organ size and hydration status, body composition including plasma volume, enzyme activity involved in metabolism, and drug excretion [2, 5, 6].

Generally, differences are observed at each stage of the PK process including absorption, distribution, metabolism, and excretion due to sex-based variations. Drug absorption occurs throughout the gastrointestinal tract, respiratory tract, and skin, and is influenced by the type of drug and the route of administration. For example, absorption of orally administered drugs is influenced by gut transit times, gastric pH, lipid solubility of the agent, hepatic metabolism, etc. [2]. Generally, females have a gastric pH about 0.5 higher than males and lower gastric acid secretion, resulting in approximately twice the gastric transit time [10]. Thus, drugs whose release depends on pH may have different absorption rates depending on gender. Therefore, drugs requiring an acidic environment for absorption, such as ketoconazole, an antibiotic, may exhibit lower bioavailability in females [11]. Additionally, as pH decreases, weakly basic drugs are more effectively absorbed, so the absorption of antidepressants, which are mostly weak bases, is significantly increased in females [12]. Since prolonged gastrointestinal transit time can decrease drug absorption [13], females may need to wait longer after meals to take drugs that should be administered on an empty stomach. Although transdermal drug absorption is relatively similar between males and females [14], hypotheses have been proposed that the bioavailability of transdermally administered drugs such as intramuscular and subcutaneous injections may differ in females due to differences in subcutaneous lipid content and skin thickness [5]. Additionally, females may have a decreased intake of inhaled aerosol drugs such as ribavirin and cyclosporine due to higher respiratory minute ventilation and lower tidal volume [15]. The activity of gastric alcohol dehydrogenase varies according to gender, leading to higher blood alcohol concentrations in females when consuming the same concentration of ethanol [16].

Once absorbed, drugs are distributed throughout the body via the circulation. Drug distribution depends on various physiological characteristics such as body composition, plasma volume, blood flow, tissue, and plasma protein binding [17].

Table 1 Physiological differences between men and women affecting drug processing, pharmacokinetics, and pharmacodynamics

Body composition	Females	Males
Fat mass	Higher	Lower
Lean body mass	Lower	Higher
Body weight	Lower	Greater
Free water	Higher	Lower
Heart rate variation	Higher	Lower
Gastric motility	Lower	Higher
Gastric acid secretion	Lower	Higher
Gastric fluid volume	Lower	Higher
Gastric alcohol dehydrogenase	Lower	Higher
Gastric glutathione activity	Higher	Lower
Small intestinal length	Shorter	Longer
Small intestinal fluid volume	Lower	Higher
Stomach acidity	Lower	Higher
Liver enzymes	Lower	Higher
Kidney excretion	Lower	Higher
Colon motility	Lower	Higher
Colonic glutathione activity	Lower	Higher
Transverse colon length	Higher	Lower
Mucosa associated colonic bacteria	Higher	Lower
Immune cell activation	Slower	Faster

*Data from Madlaet al. [4], Poleyet al. [7], Freireet al. [8], and European Commission [9]

Generally, males have higher total body water, extracellular water, intracellular water, total blood volume, and plasma volume compared to females. Therefore, hydrophilic (i.e., water-soluble) drugs such as acetaminophen and alcohol tend to have a larger volume of distribution in males, reducing drug concentration [2]. This leads to higher peak plasma levels in females. Conversely, females have a higher ratio of body fat to body weight, resulting in a higher volume of distribution of hydrophobic (i.e., lipophilic) drugs such as benzodiazepines in females, leading to longer durations of action. Organ blood flow is greater in males compared to females [17].

Drug metabolism occurs primarily in the liver, intestinal tract, lung, and kidney, and is a major factor influenced by gender [18]. Metabolism is divided into phase 1, which converts drugs into metabolites by introducing functional groups to increase polarity, and phase 2, which increases water solubility and facilitates excretion through conjugation reactions occurring in activated metabolites from phase 1 [19]. Phase 1 reactions include oxidation, reduction, and hydrolysis, while Phase 2 reactions include processes such as glucuronidation, methylation, and acetylation. The cytochrome P450 (CYP) enzymes are primarily involved in drug metabolism, and some enzymes such as CYP1A, CYP2E1, and CYP3A show definite differences

Table 2 Sex-Specific differences in the drug pharmacokinetics and pharmacodynamics in humans after drug administration

Drug class	Outcomes related with pharmacokinetics and pharmacodynamics	Effect
Nervous system agents		
Gabapentin	• Lower renal clearance in females	
Selective serotonin reuptake inhibitors (SSRI) (citalopram, dapoxetine, escitalopram, fluoxetine, fluvoxamine, paroxetine, sertraline)	• Females present higher plasma levels, probably related to sex-related activity of various CYP enzymes • Females respond better than males	• Enhanced effect in females (being the preferred therapy in females)
Tricyclic antidepressants (TCA)	• Free plasma concentrations of imipramine, clomipramine, and nortriptyline are elevated during pregnancy • Males respond better than females	• Reduced effect in females
Monoamine oxidase inhibitors	• Males respond better than females	• Reduced effect in females
Olanzapine	• Higher plasma levels and volume of distribution and lower clearance in females • More rapidly eliminated in males than in females	• Enhanced effect in females • Reduce dosage in females or increase dosage in males
Benzodiazepines	• Lower initial plasma levels due to larger volume of distribution, and possibly higher clearance, in females	
Diazepam	• Higher plasma levels of free diazepam in females • Larger distribution in females • Females experience a more significant impairment in psychomotor skills	• Reduce dosage in females
Midazolam	• Higher clearance in females	
Opioids	• Slower onset and offset of action in females • Females often report heightened pain perception and increased sensitivity to opioid receptor agonists	• Greater analgesic response in females • Males require greater dose of morphine and κ receptor agonists for comparable pain relief in females

(continued)

Table 2 (continued)

Drug class	Outcomes related with pharmacokinetics and pharmacodynamics	Effect
Paracetamol	• Lower plasma levels and higher clearance in males due to increased activity of the glucuronidation pathway • Females displayed lower clearance and volume of distribution compared with males	
Zolpidem	• Plasma levels and AUC are higher, and clearance is lower in females	• The recommended initial dose is lower in females
Cardiovascular system agents		
Angiotensin-converting enzyme inhibitors	• No mortality benefit in females with asymptomatic left ventricular systolic dysfunction	
Digoxin	• Females have higher serum digoxin concentrations due to reduced volume of distribution and lower clearance • Drug clearance increases during pregnancy • Females with heart failure have an increased risk of mortality on digoxin therapy	• Increased mortality in females • Reduce dosage in females
Isosorbide mononitrate	• Females had significantly higher serum plasma concentrations compared with males, probably due to the lower body weights in females	
Procainamide	• Plasma levels are higher in females due to a lower BMI and volume of distribution	
Statins	• Females have higher plasma levels of lovastatin and simvastatin	

(continued)

Table 2 (continued)

Drug class	Outcomes related with pharmacokinetics and pharmacodynamics	Effect
Selective beta blocking agents (e.g., metoprolol, propranolol, and atenolol)	• Females have higher plasma levels due to a smaller volume of distribution and slower clearance • Greater reduction in blood pressure and heart rate in females treated with metoprolol and propranolol	• Enhanced lowering of blood pressure and heart rate when exercising in females
Labetalol	• Labetalol concentrations are higher in females	
Verapamil	• Increased bioavailability and decreased clearance of oral verapamil in females compared with males • Females display faster clearance of verapamil after intravenous administration • Greater reduction in blood pressure and heart rate in females • Greater gut absorption in females	
Torasemide	• Higher C_{max} and lower clearance in females	
Antithrombotic agents		
Aspirin	• Bioavailability and plasma levels are higher in females possibly due to lower activity of aspirin esterase, larger volume of distribution and lower clearance in females than in males • More active in male platelets • Aspirin resistance is more frequent in females	• Higher protective effect against stroke in females • Stronger protective effect against heart attack and myocardial infarction in males
Heparin	• Females had higher plasma levels and activated partial thromboplastin time values than males due to a lower clearance • Females had increased partial thromboplastin time, even after weight-adjusted dosing, suggesting an increased sensitivity	

(continued)

Table 2 (continued)

Drug class	Outcomes related with pharmacokinetics and pharmacodynamics	Effect
Vorapaxar	• C_{max} and AUC are higher in females, but no dose adjustment is required	
Warfarin	• Higher free plasma levels in females	• Reduce dosage in females to reduce the risk of excessive anticoagulation in females
Anaesthetics		
Propofol	• Plasma propofol levels decline more rapidly in females at the end of infusion • Less sensitive in females	• Females require higher doses than males for the same effect
Lidocaine	• Females have a larger volume of distribution and may require a higher intravenous bolus dose than males • Higher free plasma levels in females receiving oral contraceptives	• Females may require higher intravenous bolus doses to achieve the same plasma levels
Corticosteroids		
Glucocorticoids (e.g., Prednisolone)	• Oral clearance and volume of distribution of prednisolone are higher in males	
Glucocorticoids (Methylprednisolone)	• Females are more sensitive to the effects of methylprednisolone	
Musculo-skeletal system agents		
Neuromuscular blocking drugs (e.g., atracurium, pancuronium, rocuronium, vecuronium)	• Lower volume of distribution, higher plasma levels, faster onset and prolonged duration in females • Females are more sensitive and require lower doses than males due to a smaller volume of distribution	• Administer higher doses to males for rapid onset of action
Respiratory system agents		
Theophylline	• Metabolism is faster and half-life is shorter in females than in males	
H1-antihistamines	• Slower metabolism and elimination in females	

*Adapted from Tamargo et al. [6]
CYP cytochrome P450 isoforms, *AUC* area under the curve, *BMI* Body mass index, *Cmax* peak plasma drug concentrations

according to gender [20]. CYP1A2 and CYP2E1 exhibit higher activity in males than in females, which can explain the higher metabolic rate of certain drugs in males [21]. Conversely, young females have been shown to have weight-normalized drug clearance rates 20–30% higher than those of young males primarily mediated by CYP3A, indicating statistically significant increases in CYP3A activity in females compared to males [21, 22]. In terms of hepatic transporters, males show higher hepatic P-glycoprotein levels compared to females, leading to lower substance concentrations of drugs in males due to the effective reduction of intracellular concentrations by P-glycoprotein, resulting in a lower drug clearance rate in males [17].

Excretion of drugs and their metabolites occurs through organs such as the kidneys, liver, and lungs. The kidneys are the primary organs responsible for excretion and sex differences have been reported in glomerular filtration, tubular secretion, and tubular reabsorption [2]. Females have a glomerular filtration rate 10–25% lower than that of males and reduced tubular secretion and reabsorption functions, resulting in lower renal clearance of drugs primarily excreted by the kidneys [11, 23]. This difference is particularly prominent in elderly females, necessitating caution when administering drugs [24]. Renal clearance of drugs that are neither actively secreted nor reabsorbed depends on glomerular filtration rate and is directly proportional to body weight. Therefore, sex-related differences in the clearance of such drugs are mainly attributable to body weight [17].

Differences in PD have not been studied as extensively as PK because of difficulties in quantifying them, such as imbalanced sex representation in clinical trials and potential variation in modulation by sex hormones, among other factors. However, prospective and retrospective analyses of clinical trials have revealed sex-specific differences in the pharmacodynamics of several widely used drugs in terms of efficacy and safety. A well-known sex difference in PD is the increased prevalence of QT interval prolongation observed in females who have taken certain drugs [4]. The withdrawal of terfenadine, astemizole, and cisapride monohydrate due to these pharmacodynamic effects was reported in a GAO report published in 2001 [1]. Additionally, females are known to have a higher incidence of drug-induced liver toxicity, gastrointestinal adverse reactions to NSAIDs, and allergic skin rash compared to males [25]. Factors such as sex hormones, differences in exposure to steroid hormones, and a higher rate of polypharmacy in females contribute to gender differences in drug responses. Sex hormones are important factors that contribute to differences in pharmacological outcomes between females and males. Endogenous steroid sex hormones naturally occur during menstrual cycles, pregnancy, and the transition to menopause, affecting the safety and efficacy of drugs [26]. In addition, females may use exogenous hormones, such as combined estrogen and progestogen for contraception or to control menopausal symptoms. These exogenous hormones can affect the pharmacokinetics and pharmacodynamics of other medications and may, in turn, be influenced by their own metabolism [26]. Physiological changes occurring during pregnancy also influence drug pharmacokinetics. These changes include increased plasma volume, stroke volume, and heart rate, compensated respiratory alkalosis, and decreased plasma albumin, which can influence drug distribution and clearance. Furthermore, sex differences in non-adherence behaviors to medications may affect drug responses. An analysis of gender patterns in medication non-adherence found that forgetfulness or intentional changes in dosing were more common causes of non-adherence in males, while non-adherence in females was often due to adverse drug reactions (ADRs) [27]. In conclusion, understanding sex-based differences in drug responses is essential for optimizing drug therapy and improving patient outcomes.

2.2 Sex and Gender-Specific Side Effects

It is known that there are sex differences in the types and magnitude of risks associated with ADRs. Women have been reported to experience approximately 1.5 to 1.7 times more ADRs than men [28]. Various sex- or gender-related factors contribute to the differences in the risk of ADRs. Sex-related factors include hormones, genetics, metabolic processes, anatomical characteristics, and organ function [29]. These biological factors influence PKs and PDs such as clearance rates and volume of distribution, leading to differences in the risk of ADRs [30]. Moreover, these can affect drug-taking patterns such as polypharmacy. Gender-related factors encompass cultural and psychological factors including gender roles, access to medical resources and opportunities, and adherence to gender norms [29]. These may include differences in lifestyle habits and attitudes towards ADR occurrences, which can also impact the reporting of ADRs and visits to healthcare facilities [6, 31] (Table 3).

Between January 1997 and December 2000, the U.S. FDA withdrew eight drugs from the market because they posed higher risks of adverse events to females, primarily cardiovascular issues such as valvular heart disease and Torsades de Pointes [1]. While the four of these drugs were prescribed more frequently to females, the other four were commonly prescribed to both sexes yet still showed a higher risk in females. Females typically exhibit prolonged QT intervals, which increases the prevalence of Torsade de Pointes as an ADR. In fact, two-thirds of drug-induced Torsade de Pointes cases occur in females [32].

Even for drugs that were not withdrawn, numerous studies that utilized spontaneous reporting data have presented an elevated risk of adverse effects for females. Yu et al. analyzed reports from the Food and Drug Administration Adverse Event Reporting System (FAERS) for 668 drugs involved in the top 20 most commonly used long-term treatment regimens in the U.S [33]. They found sex differences in all treatment regimens, with 266 drug-event combinations remaining significant after removing confounding effects. At the system organ class level, renal and urinary disorders, congenital, familial, and genetic disorders, cardiac disorders, and blood and lymphatic system disorders were more frequently reported in males.

In contrast, ear and labyrinth disorders, musculoskeletal and connective tissue disorders, and gastrointestinal disorders were more frequently reported in females. Some of these differences persisted even after adjusting for confounding effects, suggesting inherent sex differences in ADRs. A study analyzing data from the PHARMO Database Network in the Netherlands found no difference in the ratio of hospital admissions due to severe ADRs between females and males, although

Table 3 Sex-specific differences in the drug adverse events in humans after drug administration

Drug class	Sex differences in adverse events
Anti-obesity agents	
Centrally acting anti-obesity products (Phentermine, dexfenfluramine, or fenfluramine)	• Cardiac valvulopathy occurs more frequently in females
Nervous system agents	
Antipsychotics	• Females exhibit higher instances of extrapyramidal and anticholinergic effects and QTc prolongation • Males report higher frequency of sexual problems
Benzodiazepines (diazepam)	• Greater impact on impairing psychomotor skills in females • Dependency is more prevalent among females
Opioids	• Females tend to experience more nausea, vomiting, and respiratory depression, despite requiring smaller doses for pain control
Paracetamol	• Acute liver failure resulting from overdose is more common among females
Zolpidem	• Females are vulnerable to impairment in morning-after activities
Cardiovascular system agents	
Antiarrhythmic drugs	• Females have an elevated risk of QT prolongation and Torsades de Pointes
Angiotensin-converting enzyme inhibitors	• Dry cough occurs at a 2–3 times higher in females
Digoxin	• Females with heart failure experience higher mortality rates
Procainamide	• Systemic lupus erythematosus is more prevalent in females
Statins	• Older females with low body weight are more prone to myopathy
Selective beta blocking agents (e.g., metoprolol, propranolol, and atenolol)	• Females experience enhanced blood pressure and heart rate with metoprolol • It is recommended for females to maintain digoxin plasma levels below 0.8 ng/mL
Calcium channel blockers	• Females have a heightened risk of experiencing edema
Diuretics	• Females have higher rates of hospitalizations due to hypo-osmolarity, hypokalemia, and hyponatremia, as well as a heightened risk of arrhythmias
Thiazides	• Females experience higher rates of hyponatremia and hypokalemia
Antithrombotic agents	

(continued)

Table 3 (continued)

Drug class	Sex differences in adverse events
Aspirin	• Females have a heightened risk of bleeding • Males experience more ulcer complications
Heparin	• Females have a higher risk of bleeding
Warfarin	• Females experience more frequent and severe bleedings
Thrombolytics	• Females have a higher risk of bleeding and intracranial hemorrhage
Anesthetics	
Anesthetic drugs	• Females exhibit a higher susceptibility to adverse drug reactions postoperatively
Respiratory system agents	
H1-antihistamines	• Females are more susceptible to sedation and drowsiness

*Adapted from Shan et al. [31] and Tamargo et al. [6]

specific drug-ADR combinations did show sex differences [34]. In females taking anticoagulants, the risk of rectal bleeding was higher compared to males, whereas the risks of persistent hematuria, hemoptysis, and subdural hemorrhage were lower. Females taking thiazide diuretics showed higher risk of hypokalemia than males.

In examining sex-based differences in reported AEs for various drug categories, disparities have been observed across multiple AEs. An analysis of the Swedish pharmacovigilance database revealed significant sex differences in ADRs related to antihypertensives [35]. Out of ten subgroups of antihypertensives, six—ACE inhibitors (ACE-I), ACE-I with thiazide combinations, angiotensin receptor blockers (ARB) with thiazide combinations, thiazides, diuretics and potassium-sparing agents, as well as dihydropyridine calcium-channel-blockers (DHPs)—showed notably higher reports of ADRs among females. A study utilizing data from the Korea Adverse Event Reporting System (KAERS) on the AEs of zolpidem found that parasomnia—including somnambulism and paroniria—and cardiovascular disorders, such as coronary thrombosis, were more frequently reported among females [36]. In contrast, males were more likely to report cognitive impairments such as delirium, insomnia-related disorders, and movement disorders. Another research using KAERS data on antidiabetic drugs demonstrated a higher frequency of AE reports in females across various categories and drug types, particularly in gastrointestinal diseases, metabolic disturbances, and nutritional disorders [37]. A significantly higher risk of AEs has been observed in females compared to males, including urinary and genital infections with sodium-glucose cotransporter 2 inhibitors, edema with thiazolidinedione, hyperglycemia with insulin, and urinary tract infections with dipeptidyl peptidase 4 inhibitors. Additional studies investigating AEs associated with cardiovascular drugs in the KAERS data indicated that not only was the overall reporting ratio for cardiovascular drugs higher in females, but specific drugs such as beta blockers and calcium channel blockers also exhibited this trend [38]. Moreover, females reported a higher incidence of musculoskeletal disorders, gastro-intestinal system disorders, and psychiatric disorders. The most frequently reported AEs in females included edema, fracture, and arthralgia.

Recent research into the coronavirus disease (COVID-19) pandemic revealed that severity and mortality rates are generally higher in male patients [39, 40]. Studies on the COVID-19 vaccines have also shown sex differences in efficacy and ADRs. According to a report by the U.S. Centers for Disease Control and Prevention (CDC), although females represented about 61% of those vaccinated against COVID-19, they accounted for approximately 79% of AE reports [41]. Thrombotic events were more commonly observed in females vaccinated with virus vector vaccines [42]. Additionally, the risk of serious AEs after booster vaccinations was higher in females under 40 and males over 50 [43]. These findings suggest that if sex differences are not adequately considered in the development of drugs and vaccines, they could have different safety impacts on males and females.

With the increasing recognition of sex-related differences in the risk of AEs, various regulatory measures have been implemented to account for these differences to ensure drug safety. When the FDA first approved zolpidem in 1992, the label included no gender-related recommendations. However, subsequent studies showed that 8 h after taking zolpidem, 3% of males and 15% of females had blood drug concentrations exceeding the levels associated with driving impairment [44]. For the extended-release form of zolpidem, 33% of females and 25% of males exceeded these levels. Due to the high reporting of these risks, the FDA issued a safety communication in 2013, lowering the recommended doses of zolpidem in females and requiring manufacturers to include different dosing recommendations for males and females [45]. Specifying the risk of AEs and recommending proper treatment options based on sex and/or gender by regulatory agencies can enhance the safety of drugs and provide better treatment outcomes for patients.

3 Clinical Trial Regulations and Sex Representation: Regulatory Perspectives in Drug Approval

3.1 Clinical Trials and Sex and Gender Representation

To ensure efficacy and safety of pharmaceuticals, strict guidelines must be followed, including the validation of a drug's pharmacodynamics, pharmacokinetics, and pharmacological effects through cell and tissue systems, experimental animals, and human clinical trials. Clinical trials are utilized to evaluate safety and efficacy of medical products developed to prevent, treat, or diagnose a wide range of diseases and conditions. The stages of clinical trials range from preclinical studies that evaluate new drug candidates using cells and experimental animals to phase 1, 2, 3, and 4 trials involving human subjects.

Recognizing the importance of sex-specific analysis in drug development requires incorporating potential sex differences from the early stages of the process. Among the clinical factors that need to be considered regarding sex differences, women, for example tend to take more medications, increasing the risk of adverse drug interactions. They are also more sensitive to side effects and are more frequently prone to medication overdose. However, these facts are often overlooked in the drug development process with women frequently viewed as 'smaller men', thus generating scientific evidence that fails to account for sex-based variation in medications responses. The consequent knowledge gaps regarding dose, safety, and efficacy for females make it challenging for clinicians to prescribe appropriate medications and may contribute to increased AEs in women. Sex differences in PK and PD should be considered from the earliest stages of drug development and throughout all preclinical and clinical studies. This involves considering sex differences in the expression and function of target receptors in preclinical studies, as well as enrolling a sufficient number of both men and women in later phase studies. Reflecting sex consideration from the outset enables the identification of differences in drug safety and efficacy between men and women, facilitating the advances toward more personalized therapies.

Currently, there is growing awareness of the need for adequate women representation in clinical trials. However, over the past several decades, human females and female animals have been less considered in clinical research compared to males [46]. Traditionally, women and children were considered vulnerable populations requiring strong protection from research risks. Pregnant women, in particular, have been routinely excluded from clinical research due to potential direct or indirect negative effects on the fetus, such as teratogenesis or reduced placental blood flow. A 2017 study revealed that women's enrollment rates in randomized controlled trials (RCTs) were below 50% in most disease areas. For instance, only about 38% of participants in cardiovascular clinical trials registered from 2010 to 2017 were women [47, 48]. This underrepresentation is particularly significant in conditions such as cardiovascular diseases despite the prevalence rates in women are similar to or higher than in men. From 1960 to 2013, only 1.3% of pharmacokinetic studies included pregnant women, meaning that ADRs in women are less likely to be detected before post-marketing surveillance [49].

Initially, women, pregnant women, and other vulnerable groups were excluded from early clinical trials with the good intention to protect them from potential research risks. However, this approach is now increasingly recognized as depriving these populations of the right to be protected from unfounded scientific practices, rather than safeguarding them. Excluding vulnerable groups from clinical trials means that there is a lack of evidence regarding efficacy, safety, PK, and PD for these populations, leaving them more susceptible to harm. This can eventually lead to the development of drugs that are ineffective or even detrimental to the health of these overlooked populations.

The recent COVID-19 pandemic has particularly highlighted the ethical issues and healthcare implications of not enrolling women and pregnant women in clinical trials. Men and women respond differently to COVID-19 infections, with men experiencing higher morbidity and mortality, potentially due to sex-based immunological differences and gendered differences such as behavioral and psychosocial factors [50, 51]. During the development of COVID-19 vaccines and treatments, clinical data for emergency use authorizations were urgently required and accumulated at an unprecedented rate, surpassing the availability of clinical data to support timely enrollment of women and pregnant women in trials [52]. As a result, women were underrepresented in COVID-19 treatment trials and vaccine clinical studies, raising concerns that recommendations may not be suitable for women or could even harm women's health [53, 54]. The exclusion of pregnant and lactating women from early clinical trials also led to inconsistent advice on the safety of COVID-19 vaccinations from health authorities and experts worldwide [55].

As concerns rise over the underrepresentation of females in clinical trials, the importance of their participation has gained traction, prompting various countries over recent decades to recommend or mandate the inclusion of sex considerations in clinical trials. Nonetheless, challenges in achieving adequate representation of women in clinical trials persist, mainly due to ethical concerns and misinformation regarding the enrollment of pregnant women in clinical studies.

This protective approach with ethical considerations limits the autonomy of pregnant women and excludes them from research, potentially harming both the mother and the fetus due to limited evidence [55]. A shift in perspective is required to promote inclusion, as excluding them without unjustifiable reasons prevents the opportunity to assess potential risks in a controlled environment.

3.2 Clinical Trial Regulations and Sex and Gender Considerations

Thalidomide, marketed as a sedative in the late 1950s, was widely used in Europe, Canada, Australia, and other regions for several years. However, its use was banned in most countries after the thalidomide tragedy that women who took thalidomide during pregnancy experienced numerous side effects, including phocomelia, congenital heart disease, malformations of the inner and outer ear, and ocular abnormalities [56]. In response, in 1977, the U.S. FDA issued guidelines recommending the exclusion of women of childbearing potential from phase 1 and early phase 2 clinical trials [57]. This was due to the influence of the ovarian and hormonal cycles on PK and PD, which could decrease the homogeneity of the study population and complicate experimental designs [58]. It was also assumed that study results from men could be appropriately applied to women, and there were concerns about the potential risks to placental or fetal safety due to pregnancy [59]. Since key safety and dosage decisions are made in the early stages of clinical trials, failing to adequately include women or to differentiate outcomes by sex could lead to dosages based on male physiology, increasing the risk of ADRs in women. This issue stems from the persistent of underrepresentation of women in clinical trials.

Starting in 1985, the FDA required that New Drug Applications (NDAs) include data on the efficacy and safety of drugs for different sex, age, and race/ethnicity subgroups [60]. In 1988, the FDA issued guidelines on the format and content of safety and efficacy analyses in NDAs, demanding research into whether changes in PK and PD within various sex, age and race subgroups were consistent [60]. However, a 1990 GAO report revealed that the NIH's policy on including women in clinical trials was not consistently applied across institutions and was poorly communicated within the research community [61]. The GAO also found that from 1988 to 1991, women were underrepresented in the clinical trials of over 60% of new drugs approved [62].

In 1993, the U.S. FDA introduced the 'Guideline for the Study and Evaluation of Gender Differences in the Clinical Evaluation of Drugs', which encouraged the inclusion of women in phase 1 and early phase 2 trials of drugs used by both sexes [63]. This guideline acknowledged that if significant gender differences are identified early in clinical trials, subsequent phases should be designed to address such disparities. It also noted that restricting the participation of women of childbearing potential was ethically inappropriate as it limited their autonomy and decision-making capacity. Furthermore, the June 1993 National Institutes of Health (NIH) Revitalization Act (Public Law 103–43) mandated that women and minorities must be included in all clinical research, that the number of participants in phase 3 clinical trials must be sufficient to conduct valid analyses of potential sex/gender and racial/ethnic differences, and that such groups must not be excluded due to the cost of clinical trials [64]. It also required the creation of programs and supportive activities to enroll women and minorities in clinical trials [64]. However, this NIH policy did not apply to the majority of premarketing clinical research, which consists of industry-directed trial enterprises [60]. In 1998, the FDA established the 'Demographic Rule', requiring sponsors to analyze safety and efficacy data by age, race, and sex when applying for an Investigational New Drug (IND) [65]. Despite improvements in adherence to NIH guidelines, a 2000 GAO report found that analyses by sex were not being properly conducted, and protocols that mandated scientific analysis and reporting of gender differences in phase 3 and beyond were lacking [66].

In 2012, the Food and Drug Administration Safety and Innovation Act directed the inclusion of sex and other demographic subgroups in clinical trials and the monitoring of sex-disaggregated safety and efficacy data [67]. The legislation also included monitoring of how subgroup analyses based on sex, age, race, and ethnicity were presented in NDAs, and ensuring that data on safety and efficacy for each subgroup were provided to the public in a timely manner. In 2014, the FDA Action Plan was introduced to enhance the collection and reporting transparency of demographic subgroup data and to improve the completeness and quality of analyses [68]. A 2015 GAO report indicated that clinical trial organizations often did not report summary data that could identify potential sex differences, and the NIH lacked fields indicating whether research included plans for analysis by sex [69]. In 2015, the NIH issued regulations considering sex as a biological variable, requiring the inclusion of both men and women in clinical trials and considering sex throughout the development of research questions, study design, data collection, analysis, and reporting [70]. In

2020, the FDA clarified the differences between the terms 'sex' and 'gender' in its documents and provided guidance on naming them as separate subgroups [71]. In 2022, the U.S. House of Representatives had passed a bill to increase the diversity of populations enrolled in clinical trials for new drugs. This legislation requires sponsors to submit study enrollment goals for demographic groups for phase 3 or pivotal studies and plans to achieve those diversity [72].

In Canada, Health Canada initiated the inclusion of women in clinical trials in 1997 with the 'Guidance Document on the Inclusion of Women in Clinical Trials' [73]. These guidelines called for the inclusion of not only premenopausal women but also postmenopausal women in all stages of drug development, along with the subgroup analysis by sex. However, this was not an absolute requirement, but rather a recommendation to encourage women's participation in clinical trials. Revised in 2013, the guideline recommended that trial sponsors of clinical trials applications for drugs to be used by both men and women should include both sexes in nonclinical studies and clinical trials to identify potential sex-related differences in efficacy and safety [74]. Health Canada established the Scientific Advisory Committee on Women's Health Products (SAC-HPW) in 2019 to provide timely scientific advice related to women's health, and integrated Sex- and Gender-Based Analysis Plus (SGBA Plus) into regulatory processes and practices to improve the safety of drugs for women [75, 76]. In October 2022, Health Canada initiated a procedure requesting that clinical trial data segmented by sex and other subpopulations for safety and efficacy be included in drug submissions [77].

In Europe, clinical research primarily involved men until the 1990s. However, the International Council for Harmonization (ICH), formerly the International Conference on Harmonization, mandated that women achieve representation in trial populations, which is generally possible by reflecting the prevalence of the studied disease or condition by gender [78]. The ICH guidelines also required reporting demographic characteristics of the patient population, including gender, and if the sample size was sufficient, presenting analysis results by specific populations to explore differences based on demographic characteristics. When including women in clinical trials, it was recommended to exclude pregnant women and ensure that women of childbearing potential use effective contraception [79]. However, due to the lack of reinforcement of these recommendations and the softness of statements requiring adequate sample sizes, reporting of gender distribution and subgroup analysis results has generally not been well executed [78]. In 2014, the European Medicines Agency (EMA) issued regulations across the European Union for conducting clinical trials, establishing efficient procedures for the authorization process for a clinical trial, protection of subjects, and safety reporting [80]. This document included requirements that subgroups, including age and gender, be represented in the population using the medicinal product and that all exclusion criteria be justified. It also specified that clinical trial data must be segregated by sex and gender. In 2019, the EMA published various principles and evaluation strategies recommending subgroup analysis to estimate treatment effects and to compare them among subgroups [81]. In Europe, unlike other countries such as the U.S., the term 'gender' was used instead of biological sex.

Efforts to include minorities in clinical trials have heightened awareness of sex differences, leading to recommendations for the inclusion of pregnant women in clinical trials. Despite about half of pregnant women being prescribed drugs for both obstetric and non-obstetric indications during pregnancy [82], pregnant and lactating women have been almost entirely excluded from clinical trials. Estimating outcomes for pregnant women based on studies conducted on men and non-pregnant women can be potentially hazardous due to significant differences in immune responses and dynamics during pregnancy. Thus, there is often a lack of information to assess the benefits of drug use and the risks, such as teratogenicity, for both the pregnant woman and the fetus. In such situations, clinicians must judge the appropriate dosage when prescribing medications to pregnant women [83]. Limited or absent information can lead to safety issues and, even if not, may deter necessary treatment or cause hesitation among clinicians. However, not providing appropriate treatment for pregnant women can lead to even more adverse outcomes. For example, untreated depression in pregnant women can lead to maternal suicidal ideation, preterm delivery, and small-for-gestational-age outcomes [84]. To address these knowledge gaps and mitigate potential negative impacts on patient care, it is crucial to enroll pregnant individuals in clinical trials and collect the necessary data to support medication use during pregnancy as early as possible.

The inclusion of pregnant women in clinical trials is primarily aimed at supporting the development of treatments for pregnancy-related conditions such as preeclampsia and intrahepatic cholestasis of pregnancy, as well as for diseases that occur with pregnancy, such as hypertension and diabetes mellitus [55]. In many cases, sponsors first conduct efficacy trials on men and non-pregnant women and if initial studies demonstrate efficacy, they proceed with so-called bridging studies that include pregnant women [52]. Furthermore, ongoing efforts are being made to enhance awareness of the inclusion of pregnant individuals in clinical trials to bridge clinical knowledge gaps and encourage timely evidence generation for pregnant women. The U.S. FDA supports research on women and the perinatal population through the Office of Women's Health (OWH) and the FDA Perinatal Health Center of Excellence (PHCE), publishes guidance on including pregnant women in clinical trials [85], and collects tools and data for the treatment of mothers and infants through the Maternal and Pediatric Precision in Therapeutics (MPRINT) Hub. Moreover, the Task Force on Research Specific to Pregnant Women and Lactating Women (PRGLAC) publishes recommendations and implementation plans for improving research on therapeutics for pregnant and lactating women [86]. The NIH strongly encourages the inclusion of pregnant women in clinical research in all situations where it is scientifically valid and ethically permissible [87].

During the development of the COVID-19 vaccine, the World Health Organization (WHO) emphasized the need for data related to pregnancy, highlighting the necessity of generating pregnancy-related safety data from participants who inadvertently become pregnant during bridging studies and phase 3 clinical trials. WHO also announced that vaccine developers and funders should prioritize evaluating the safety and immunogenicity of vaccines in pregnant women during development and prioritize assessing safety and effectiveness in post-marketing surveillance [88].

3.3 Sex and Gender Considerations at Each Stage of Clinical Trials

To consider sex at each stage of drug clinical trials, prudent selection of study samples, transparent reporting of results by sex, and pertinent approaches tailored to the characteristics of each trial stage are necessary. Table 4 represents modified recommendations for considering sex in drug development stages proposed by Tannenbaum et al. [89].

Table 4 Recommendations for considering sex/gender in pre-marketing drug clinical trials

Stage		Recommendations for considering sex
Preclinical trials	In vitro experiments	1. Reporting the genetic sex of human or animal cell lines and cultured cells 2. Comparing cells from males and females and reporting the results 3. Adding gonadal hormones to plated cells to confirm effects
	Animal experiments	1. Reporting the sex of animals included in the study 2. Including both male and female animals in animal experiments 3. Including pregnant animals 4. Reporting results separately for male and female animals 5. Exploring sex differences using small stratified samples 6. Independently evaluating the impact of sex 7. Considering the role of gonadal hormones
Phase 1 and 2 trials		1. Including equal numbers of males and females in phase 1 and 2 trials, as well as bioequivalence studies 2. Collecting data on premenopausal age, menopause, pregnancy and negative pregnancy events, ovariectomy, hormonal contraceptive use, and hormone replacement therapy 3. Controlling for menstrual cycle phase when testing premenopausal women 4. Determining if sex specific criteria for biomarkers are needed 5. Investigating pharmacogenomic mechanisms to explain sex differences 6. Reporting results categorized by sex
Phase 3 drug trials		1. Calculating sample size in advance based on preliminary data to detect differences between or within sex subgroups 2. Using targeted methods to recruit and retain women 3. Including pregnant women in clinical trials 4. Collecting data on premenopausal age, menopause, pregnancy and negative pregnancy events, ovariectomy, hormonal contraceptive use, and hormone replacement therapy 5. Considering adaptive clinical trial designs to determine doses based on sex 6. Reporting results categorized by sex

*Adapted from Tannenbaum et al. [89]

In preclinical studies, both cell-based and animal-based in vivo experiments should take into account sex when evaluating the toxicity and efficacy of candidate substances. However, historically, these studies have primarily focused on male cells and male animals, leading to an oversight of differences between males and females. A study examining publications from 2011 to 2012 found that 22% of animal research did not report sex, and among those that did, 80% included only males. Similarly, 76% of cell research papers did not report gender, and among those that did, over 70% included only males [90]. Even in investigations of 22 drugs approved between 2011 and 2015, only 9% of preclinical PD studies included female animals [91]. Preclinical studies involving animals primarily aim to provide pharmacological dose–response effects rather than identifying differences between females and males. There are also opposing views on including female animals in preclinical studies due to reasons such as research costs and time constraints. Nonetheless, understanding female physiology and pathophysiology as comprehensively as that of males is necessary to avoid unexpected AEs that may lead to the withdrawal of drugs from the market [92].

Ideally, both male and female cells should be used in experiments during the early stages of drug development to identify sex differences. While cellular responses to compounds or biological agents in the human body may be regulated by hormones or metabolism, sex differences may not be evident when cells are present under identical and static culture conditions in vitro. Therefore, adding gonadal hormones to cell cultures has been proposed to evaluate sex differences [93]. To investigate sex effects in animal experiments, both males and females should be included, and results should be separately analyzed by sex. Excluding female animals is typically done to control for hormonal fluctuations, but the influence of these hormonal changes should also be considered [94]. Additionally, a 2×2 factorial design, which evaluates both sex and other variables simultaneously and assesses the interaction between sex and other variables, can be considered. This design presents the main effects of sex and other independent variables on outcome measurements and examines whether one independent variable modifies the effects of another variable [95]. When studying sex differences in metabolism in preclinical research, it is suggested to investigate the influence of estrous cycle fluctuations, which can be altered by rapid changes in neuronal firing, neuropeptide secretion, and autonomic nervous activation [93]. Furthermore, sex differences in acute stress and pain responses mediated by different hormonal systems should be considered. As sex differences may arise due to various factors such as hormones, metabolites, neural inputs, and body composition in the in vivo and in vitro environment, tools to study these differences are necessary [93].

In Phase 1 studies, researchers primarily test drugs on a small number of healthy subjects to collect PK data and information on interactions with the human body. Phase 2 studies confirm the efficacy and toxicity of the drug in hundreds of subjects with the target disease [96]. Most bioequivalence studies have traditionally been conducted on males under the assumption that PK are similar between sexes, which is not accurate. Therefore, to identify meaningful gender differences, it is recommended to target half of the subjects in Phase 1 and 2 trials and bioequivalence studies to be female participants. Essential information related to menstruation and hormone replacement therapy that may affect drug exposure and response should be collected. Significant sex differences have been observed in various categories of biomarkers, including lipids, adipokines, inflammation, endothelial dysfunction, and kidney function [97]. Therefore, applying sex-specific biomarkers may enhance insights into differential responses. Reporting data by sex and exploring differences in PK, PD, and immunogenicity based on sex are recommended.

Phase 3 clinical trials aim to demonstrate the efficacy and additional safety of a drug in a specific population, conducted on a larger scale and for a longer duration than previous trials [96]. However, women representation in Phase 3 trials remains low. A study found that in about 26% of publicly available registration dossiers of FDA-approved and frequently prescribed drugs, there was a difference of more than 20% between the proportion of women with the disease and the proportion of women who participated in the clinical trials [98]. Another study found that women were underrepresented in Phase 3 clinical trials for drugs approved in Europe between 2011 and 2015, particularly in medications for schizophrenia, hypercholesterolemia, and heart failure [91].

To detect sex differences in Phase 3 clinical trials, it is essential to calculate the necessary sample size to detect minimal clinically important differences in drug efficacy or safety based on previous clinical trial and simulation results. Encouraging participation of pregnant women and developing effective communication strategies to explain the importance of their participation in clinical trials are essential to recruit participants to meet the calculated sample size. To promote the involvement of these target patients, measures to reduce physical, economic, and social barriers to clinical trial participation can also be implemented, such as providing transportation support, flexible scheduling, and online participation options [46]. Data on menstruation, pregnancy, and hormone therapy should also be collected in Phase 3 trials as in Phase 2 trials. All trial results should be analyzed and reported by sex to understand the impact of sex difference on outcomes.

4 Pharmacovigilance Through a Gender Lens

4.1 Approaches to Monitoring and Reporting in Pharmacovigilance Database

Pharmacovigilance is a branch of pharmacological science encompassing the detection, assessment, understanding, and prevention of adverse effects or other drug-related problems associated with pharmaceutical products [99]. It primarily involves collecting and monitoring information on AEs occurring during medication use from healthcare providers and patients. Safety information collected through pre-marketing clinical trials is limited in detecting rare or long-term safety issues and ADRs specific to populations not well represented in clinical trials. Therefore, close monitoring is required for rare ADRs, ADRs with a long latency, and ADRs in specific populations in the post-marketing phase [100].

One widely used method in pharmacovigilance is the spontaneous reporting system (SRS), where patients, healthcare professionals, or caregivers can report suspected ADRs following drug use. These reports are collected and analyzed to detect signals not identified during pre-marketing clinical trials [101]. Several studies, as outlined in Sect. 2.2 of the current text, have reported variations in reporting frequency and severity of ADRs based on sex and/or gender. Investigating sex differences using AE reporting data can reveal safety disparities not identified during pre-marketing clinical trials.

However, accurately interpreting physiological differences by sex and/or gender in SRS data requires considering various factors that may influence AE reporting [102]. Firstly, women tend to utilize healthcare services at higher rates than men, leading to higher exposure to prescription drugs [103, 104]. Higher visit frequencies due to reproductive health needs contribute to this phenomenon. Consequently, even if there are no differences in ADR occurrence between women and men, the higher exposure of women to prescription drugs may lead to more AE reports [105]. Secondly, biases and discrimination in diagnosing specific conditions and prescribing certain drugs based on an individual's sex and/or gender by healthcare providers could affect AE onset. For example, women may receive less adequate pain management compared to men due to perceptions of higher pain sensitivity, resulting in fewer prescriptions for long-lasting opioids. Conversely, women may receive more prescriptions for antidepressants, anticonvulsants, and muscle relaxants than men [106]. When diagnosis and treatment does not adequately account for sex disparities, it may influence AE onset. Thirdly, sex and/or gender can influence how subjective experiences or events are perceived as AEs. Factors related to appearance, such as weight gain, are perceived more negatively in women than men [107], indicating a higher likelihood of reporting weight gain-related AEs in women. Conversely, men tend to be more sensitive to diseases related to sexual function [102]. Women may have fears of stigma and acceptance of reproductive/sexual problems related to AE reporting [108]. Therefore, caution is needed in interpreting AEs with different reporting possibilities based on sex and/or gender. Lastly, considering differences in health status due to environmental factors by sex and/or gender is essential. Gender interacts with socioeconomic status, increasing the likelihood of exposure to poverty, partner violence, and sexual assault in women compared to men, which can increase the risk of mental and physical health problems [109]. These factors may influence the chance of being diagnosed with a disease or recognizing AE onset.

Previous studies have suggested that these various factors may influence the following stages of AE reporting: (1) having a condition requiring medical services, (2) diagnosis and prescription, (3) identifying AE onset, and (4) AE reporting [102]. Therefore, it is necessary to consider these factors when analyzing safety signal by sex and/or gender in SRS database. While gender should be given significant consideration in drug surveillance using ADR reporting, variables such as age, race/ethnicity, which may also influence AE reporting, may be inconsistently or inaccurately included in drug surveillance databases. Therefore, these aspects should be taken into careful consideration when interpreting analysis results.

4.2 Sex and Gender-Specific Drug Safety Study

When considering sex differences in pre-marketing clinical trials of drugs, the primary focus is on including women and presenting results based on sex and/or gender. In contrast, post-marketing clinical studies should evaluate the effectiveness and safety by sex and/or gender in real-world clinical settings, using large-scale real-world data that includes women and special populations such as pregnant women, the elderly, and children who are underrepresented in clinical trials. Recommended steps for conducting post-marketing clinical studies considering sex and/or gender have been proposed from literature review to reporting [89, 110–112] (Table 5).

Firstly, understanding the biological and epidemiological relationship between the target drug and AEs with regard to sex and/or gender is crucial to conduct research systematically. Known sex and/or gender differences in disease incidence, prevalence, and survival rates should be considered, along with specific exposure periods highly relevant to sex and gender throughout the lifespan, such as fetal and childhood development, puberty, reproductive events, and aging [113, 114]. In the literature review stage, searching for literature using keywords such as sex, gender, female, male, woman, man, girl, and boy to investigate the impact of sex and gender on the research topic and collecting relevant data is essential. A sufficient literature review should be performed to provide clear explanations in already known sex and/or gender differences in epidemiological studies, risk factors, diseases, or treatment effects.

In the research design phase, planning to ensure fair impact of sex-related disparities on the health of various subgroups is necessary. Strong scientific justification must be provided if only one sex is used or if the gender balance is not achieved in the study. This applies, in particular, when the disease or the study population is limited to one sex (e.g., ovarian and prostate cancers), when the sample sizes are too limited to maintain statistical power, or when the validity of a study using only one sex has been proven through previous literature, preliminary experimental results, or other data. Clear inclusion and exclusion criteria considering sex and/or gender of diverse individuals should be established. Balanced random allocation stratified by sex can be achieved even if the number of men and women differs [112].

Adequate sample sizes should be set to ensure statistical power of the study to test interactions between sex and exposure or outcome. However, even if the study size is too small, it can still contribute to hypothesis generation or meta-analysis. Accessibility to symptoms and diagnostic test results that account for detailed differences in the manifestation of diseases by sex should be confirmed. Additionally, it is important to document factors that may contribute to differences due to PK and PD variations between females and males, such as hormonal use, chemical exposures. Furthermore, relevant data collection tools should adequately describe both quantitative and qualitative data. Factors such as smoking, alcohol consumption, physical activity, socioeconomic factors, which may differ by sex and/or gender and may act as confounders, should be collected for analysis.

Table 5 Recommendations for considering sex/gender in post-marketing drug clinical trials

Stage	Best practices
Literature review	• Explicit identification of any documented disparities between sexes and/or genders in the epidemiology, risk factors, conditions, diseases, or treatment effects being investigated • Review of relevant literature describing documented variances, if any, between sexes and/or genders within the scope of the research area (e.g., age, ethnicity, income, occupation, and other significant social determinants of health) • Take into account established sex-based differences in disease incidence, prevalence, and survival rates • Examine the timing of exposure that may interact with sex and/or gender during specific developmental stages from a life course perspective
Research question	• Identify any sex and/or gender differences in the intervention, treatment, or outcomes being studied • Examine whether there are no sex and/or gender differences in the intervention, treatment, or outcomes under study • Consider sex and/or gender as a confounder or interaction variable while testing study hypotheses • Include gender and/or sex, or relevant groups or phenomena in research questions (e.g., differences between males and females, differences among women, gendered phenomenon such as masculinity)
Study design and methods	• Provide scientifically valid rationale for proposing a study involving only one sex, if applicable <Subject Selection> • Consider inclusion and exclusion criteria that cover sex and/or gender across diverse populations • Propose recruitment strategies designed to attain necessary sample sizes of eligible population • Consider sex-specific age-related disease incidence • Examine reproductive stages and cycles, especially when they may alter the effects of the exposure or outcome • Assess the influence of gendered social environments that could interact with the exposure or outcome <Randomization> • Apply stratified randomization by sex and/or gender to ensure balance, even if different numbers of males and females are included < Sample Size > • Ensure large enough sample sizes to test interactions between sex and/or gender and the exposure • Beware of the risk of false-negatives in underpowered sex strata • Even studies too small to detect main effects of sex can provide sex-specific data to generate hypotheses or contribute to meta-analyses of sex and/or gender differences • Conduct big data studies to avoid the risk of false-positives <Data Collection> • Consider whether exposures have the same implications in both sexes and genders • Be mindful of sex and gender disparities in pharmacokinetics and pharmacodynamics • Collect data on exogenous hormones (e.g., contraceptives, menopausal hormone therapy, testosterone, and other steroid use) and data on the reproductive cycle (e.g., follicular/luteal) and stage (e.g., prepuberty, puberty, pregnancy, lactation, pre-menopause, and post-menopause) • Collect data on covariates that may differ by sex and gender within the study population • Describe the data collection tools concerning the capture of sex- and gender-related variables of interest
Analysis and reporting	• Provide a description of the data analysis plan • Explore intermediate pathway variables to comprehend apparent sex differences • Present sample size calculations to demonstrate adequate power for testing hypotheses regarding sex and/or gender differences • Address confounding by factors associated with sex and/or gender • Interpret apparent sex and gender differences within the context of biological plausibility and social circumstances

* Data from Canadian Institutes of Health Research [110], Heidari et al. [111], Rich-Edwards et al. [112], and Tannenbaum et al. [89]

In the analysis and reporting phase, methods such as stratified analysis and pathway modeling based on sex and/or gender can be utilized [112]. When necessary, sex-related factors can be considered as confounding variables or interaction terms. Sample sizes appropriate for analyzing sex differences should be indicated in sex-based analysis. Besides statistically significant results, non-significant findings in terms of sex and/or gender aspects should also be reported. This is because even without statistical significance, if results showing a specific trend due to sex differences are repeatedly observed, they can be used in meta-analyses of multiple studies or provide information for future studies. It should be explicitly stated that sex-specific effects may be influenced by age or other biological variables.

5 Conclusion

In the field of drug safety, over the past several decades, various countries have come up with recommendations to enhance diversity across all phases of clinical trials including the inclusion of women and the analysis of safety and efficacy data stratified by sex to account for sex differences. Despite growing recognition of the importance of incorporating sex-specific perspectives, inequalities persist. Effective recruitment strategies and targeted communication need to be implemented to promote gender diversity and ensure the participation and retention of special populations, such as pregnant women, in clinical trials. Looking ahead, regulatory agencies should move beyond recommendations and require mandatory sex-disaggregated data reporting for all drug trial results submitted by the pharmaceutical industry. Continuous updates to comprehensive guidelines are necessary to account for sex disparities related not only to biological differences but also to potential sociocultural influences on drug prescription and ADRs. Transparency in reporting sex- specific data and research findings should be encouraged. For researchers involved in clinical trials across the drug lifecycle, providing education on the importance of sex- and gender-sensitive perspectives and organizing workshops to share latest research trends and findings on sex differences in drug safety could also be suggested. As awareness of sex differences in health research advances, it is essential to consider these differences at every stage of drug development to ensure better health outcomes for both men and women. Ultimately, this approach can lead to more personalized and effective treatments for all patients.

References

1. U.S. General Accounting Office. Drug safety: most drugs withdrawn in recent years had greater health risks for women; 2001. https://www.gao.gov/products/gao-01-286r. Accessed Jun 30 2025.
2. Soldin OP, Chung SH, Mattison DR. Sex differences in drug disposition. J Biomed Biotechnol. 2011;2011: 187103.
3. Messing K, Mager SJ. Sex, gender and women's occupational health: the importance of considering mechanism. Environ Res. 2006;101:149–62.
4. Madla CM, Gavins FKH, Merchant HA, Orlu M, Murdan S, Basit AW. Let's talk about sex: differences in drug therapy in males and females. Adv Drug Deliv Rev. 2021;175: 113804.
5. Soldin OP, Mattison DR. Sex differences in pharmacokinetics and pharmacodynamics. Clin Pharmacokinet. 2009;48:143–57.
6. Tamargo J, Rosano G, Walther T, Duarte J, Niessner A, Kaski JC, et al. Gender differences in the effects of cardiovascular drugs. Eur Heart J Cardiovasc Pharmacother. 2017;3:163–82.
7. Poley M, Chen G, Sharf-Pauker N, Avital A, Kaduri M, Sela M, et al. Sex-based differences in the biodistribution of nanoparticles and their effect on hormonal, immune, and metabolic function. Adv Nanobiom Res. 2022;2:2200089.
8. Freire AC, Basit AW, Choudhary R, Piong CW, Merchant HA. Does sex matter? The influence of gender on gastrointestinal physiology and drug delivery. Int J Pharm. 2011;415:15–28.
9. European Commission. Gendered innovations 2: how inclusive analysis contributes to research and innovation: policy review. Publications Office of the European Union Luxembourg; 2020. https://doi.org/10.2777/316197. Accessed Jun 30 2025.
10. Stephen AM, Wiggins HS, Englyst HN, Cole TJ, Wayman BJ, Cummings JH. The effect of age, sex and level of intake of dietary fibre from wheat on large-bowel function in thirty healthy subjects. Br J Nutr. 1986;56:349–61.
11. Whitley H, Lindsey W. Sex-based differences in drug activity. Am Fam Physician. 2009;80:1254–8.
12. Hutson WR, Roehrkasse RL, Wald A. Influence of gender and menopause on gastric emptying and motility. Gastroenterology. 1989;96:11–7.
13. Fleisher D, Li C, Zhou Y, Pao LH, Karim A. Drug, meal and formulation interactions influencing drug absorption after oral administration. Clinical implications. Clin Pharmacokinet. 1999;36:233–54.
14. Schwartz JB. The current state of knowledge on age, sex, and their interactions on clinical pharmacology. Clin Pharmacol Ther. 2007;82:87–96.
15. Nicolas JM, Espie P, Molimard M. Gender and interindividual variability in pharmacokinetics. Drug Metab Rev. 2009;41:408–21.
16. Baraona E, Abittan CS, Dohmen K, Moretti M, Pozzato G, Chayes ZW, et al. Gender differences in pharmacokinetics of alcohol. Alcohol Clin Exp Res. 2001;25:502–7.
17. Gandhi M, Aweeka F, Greenblatt RM, Blaschke TF. Sex differences in pharmacokinetics and pharmacodynamics. Annu Rev Pharmacol Toxicol. 2004;44:499–523.
18. Waxman DJ, Holloway MG. Sex differences in the expression of hepatic drug metabolizing enzymes. Mol Pharmacol. 2009;76:215–28.
19. Meyer UA. Overview of enzymes of drug metabolism. J Pharmacokinet Biopharm. 1996;24:449–59.
20. Yang L, Li Y, Hong H, Chang CW, Guo LW, Lyn-Cook B, et al. Sex differences in the expression of drug-metabolizing and transporter genes in human liver. J Drug Metab Toxicol. 2012;3:1000119.
21. Scandlyn MJ, Stuart EC, Rosengren RJ. Sex-specific differences in CYP450 isoforms in humans. Expert Opin Drug Metab Toxicol. 2008;4:413–24.
22. Greenblatt DJ, von Moltke LL. Gender has a small but statistically significant effect on clearance of CYP3A substrate drugs. J Clin Pharmacol. 2008;48:1350–5.

23. Gaudry SE, Sitar DS, Smyth DD, McKenzie JK, Aoki FY. Gender and age as factors in the inhibition of renal clearance of amantadine by quinine and quinidine. Clin Pharmacol Ther. 1993;54:23–7.
24. Hilmer S, Rochon P. Sex and age differences in geriatric pharmacotherapy. Public Policy Aging Rep. 2023;33:132–5.
25. Anderson GD. Gender differences in pharmacological response. Int Rev Neurobiol. 2008;83:1–10.
26. Moyer AM, Matey ET, Miller VM. Individualized medicine: Sex, hormones, genetics, and adverse drug reactions. Pharmacol Res Perspect. 2019;7: e00541.
27. Thunander Sundbom L, Bingefors K. Women and men report different behaviours in, and reasons for medication non-adherence: a nationwide Swedish survey. Pharm Pract (Granada). 2012;10:207–21.
28. Rademaker M. Do women have more adverse drug reactions? Am J Clin Dermatol. 2001;2:349–51.
29. Brabete AC, Greaves L, Maximos M, Huber E, Li A, Le ML. A sex- and gender-based analysis of adverse drug reactions: a scoping review of pharmacovigilance databases. Pharmaceuticals (Basel). 2022;15:298.
30. Zucker I, Prendergast BJ. Sex differences in pharmacokinetics predict adverse drug reactions in women. Biol Sex Differ. 2020;11:32.
31. Shan Y, Cheung L, Zhou Y, Huang Y, Huang RS. A systematic review on sex differences in adverse drug reactions related to psychotropic, cardiovascular, and analgesic medications. Front Pharmacol. 2023;14:1096366.
32. Drici MD, Clement N. Is gender a risk factor for adverse drug reactions? The example of drug-induced long QT syndrome. Drug Saf. 2001;24:575–85.
33. Yu Y, Chen J, Li DC, Wang LW, Wang W, Liu HF. Systematic analysis of adverse event reports for sex differences in adverse drug events. Sci Rep. 2016;6:24955.
34. Hendriksen LC, van der Linden PD, Lagro-Janssen ALM, van den Bemt PMLA, Siiskonen SJ, Teichert M, et al. Sex differences associated with adverse drug reactions resulting in hospital admissions. Biol Sex Differ. 2021;12:34.
35. Rydberg DM, Mejyr S, Loikas D, Schenck-Gustafsson K, von Euler M, Malmström RE. Sex differences in spontaneous reports on adverse drug events for common antihypertensive drugs. Eur J Clin Pharmacol. 2018;74:1165–73.
36. Joung KI. Gender differences in spontaneous adverse event reports associated with zolpidem in South Korea, 2015–2019. Front Pharmacol. 2023;14:1256245.
37. Joung KI, Jung GW, Park HH, Lee H, Park SH, Shin JY. Gender differences in adverse event reports associated with antidiabetic drugs. Sci Rep. 2020;10:17545.
38. Park HH, Kim JH, Yoon D, Lee H, Shin JY. Gender differences in the adverse events associated with cardiovascular drugs in a spontaneous reporting system in South Korea. Int J Clin Pharm. 2021;43:1036–44.
39. Jin JM, Bai P, He W, Wu F, Liu XF, Han DM, et al. Gender differences in patients with COVID-19: focus on severity and mortality. Front Public Health. 2020;8:152.
40. Kragholm K, Andersen MP, Gerds TA, Butt JH, Ostergaard L, Polcwiartek C, et al. Association between male sex and outcomes of coronavirus disease 2019 (COVID-19)-A Danish Nationwide, register-based study. Clin Infect Dis. 2021;73:E4025–30.
41. Gee J, Marques P, Su J, Calvert GM, Liu RL, Myers T, et al. First month of COVID-19 vaccine safety monitoring—United States, December 14, 2020–January 13, 2021. MMWR Morb Mortal Wkly Rep. 2021;70:283–8.
42. Jensen A, Stromme M, Moyassari S, Chadha AS, Tartaglia MC, Szoeke C, et al. COVID-19 vaccines: considering sex differences in efficacy and safety. Contemp Clin Trials. 2022;115: 106700.
43. Janekrongtham C, Salazar M, Doung-Ngern P. Sex differences in serious adverse events reported following booster doses of COVID-19 vaccination in Thailand: a countrywide nested unmatched case-control study. Vaccines (Basel). 2023;11:1772.

44. Farkas RH, Unger EF, Temple R. Zolpidem and driving impairment–identifying persons at risk. N Engl J Med. 2013;369:689–91.
45. Food Drug Administration. Drug safety communication: FDA approves new label changes and dosing for zolpidem products and a recommendation to avoid driving the day after using Ambien CR; 2013. http://www.fda.gov/Drugs/DrugSafety/ucm352085.htm. Accessed Jun 30 2025.
46. Liu KA, Mager NA. Women's involvement in clinical trials: historical perspective and future implications. Pharm Pract (Granada). 2016;14:708.
47. Daitch V, Turjeman A, Poran I, Tau N, Ayalon-Dangur I, Nashashibi J, et al. Underrepresentation of women in randomized controlled trials: a systematic review and meta-analysis. Trials. 2022;23:1038.
48. Jin XR, Chandramouli C, Allocco B, Gong EY, Lam CSP, Yan LJL. Women's participation in cardiovascular clinical trials from 2010 to 2017. Circulation. 2020;141:540–8.
49. McCormack SA, Best BM. Obstetric pharmacokinetic dosing studies are urgently needed. Front Pediatr. 2014;2:9.
50. Bischof E, Wolfe J, Klein SL. Clinical trials for COVID-19 should include sex as a variable. J Clin Invest. 2020;130:3350–2.
51. Wenham C, Smith J, Morgan R, Grp GCW. COVID-19: the gendered impacts of the outbreak. Lancet. 2020;395:846–8.
52. Modi N, Ayres-de-Campos D, Bancalari E, Benders M, Briana D, Di Renzo GC, et al. Equity in coronavirus disease 2019 vaccine development and deployment. Am J Obstet Gynecol. 2021;224:423–7.
53. Xiao H, Vaidya R, Liu F, Chang XM, Xia XQ, Unger JM. Sex, racial, and ethnic representation in COVID-19 clinical trials a systematic review and meta-analysis. JAMA Intern Med. 2023;183:50–60.
54. Stillwell RC. Exclusion of women from COVID-19 studies harms women's health and slows our response to pandemics. Biol Sex Differ. 2022;13:27.
55. Sewell CA, Sheehan SM, Gill MS, Henry LM, Bucci-Rechtweg C, Gyamfi-Bannerman C, et al. Scientific, ethical, and legal considerations for the inclusion of pregnant people in clinical trials. Am J Obstet Gynecol. 2022;227:805–11.
56. Kim JH, Scialli AR. Thalidomide: The tragedy of birth defects and the effective treatment of disease. Toxicol Sci. 2011;122:1–6.
57. U.S. Food Drug Administration. General considerations for the clinical evaluation of drugs: U.S. Department of health, education, and welfare, public health service, food and drug administration; 1977.
58. Beery AK, Zucker I. Sex bias in neuroscience and biomedical research. Neurosci Biobehav Rev. 2011;35:565–72.
59. Institute of Medicine Committee on E. Legal issues relating to the inclusion of women in clinical S. In: Mastroianni AC, Faden R, Federman D, editors. Women and health research: ethical and legal issues of including women in clinical studies, vol. I. Washington (DC): National Academies Press (US). Copyright 1994 by the National Academy of Sciences. All rights reserved; 1994.
60. Hwang TJ, Brawley OW. New federal incentives for diversity in clinical trials. N Engl J Med. 2022;387:1347–9.
61. U.S. Food Drug Administration. Guideline for the format and content of the clinical and statistical sections of new drug applications: Rockville, MD: FDA, US Department of Health and Human Services; 1988.
62. U.S. General Accounting Office. National institutes of health: problems in implementing policy on women in study populations; 1990. https://www.gao.gov/products/t-hrd-90-38. Accessed Jun 30 2025.
63. U.S. General Accounting Office. Women's health: FDA needs to ensure more study of gender differences in prescription drug testing; 1992. https://www.gao.gov/products/hrd-93-17. Accessed Jun 30 2025.

64. U.S. Food Drug Administration. Guideline for the study and evaluation of gender differences in the clinical evaluation of drugs. Federal Register. 1993;58:39406–16.
65. National Institutes of Health Revitalization Act of 1993: act to amend the Public Health Service Act to revise and extend the programs of the National Institutes of Health, and for other purposes; 1993.
66. Office of the Federal Register National Archives and Records A. 63 FR 6854—Investigational new drug applications and new drug applications; 1998. https://www.govinfo.gov/app/details/FR-1998-02-11/98-3422. Accessed Jun 30 2025.
67. U.S. General Accounting Office. Women's health: NIH has increased its efforts to include women in research; 2000. https://www.gao.gov/products/hehs-00-96. Accessed 30 Jun 2025.
68. U.S. Food Drug Administration. FDASIA Section 907: inclusion of demographic subgroups in clinical trials. https://www.fda.gov/regulatory-information/food-and-drug-administration-safety-and-innovation-act-fdasia/fdasia-section-907-inclusion-demographic-subgroups-cli nical-trials. Accessed 30 Jun 2025.
69. U.S. Food Drug Administration. FDA action plan to enhance the collection and availability of demographic subgroup data; 2014.
70. U.S. General Accounting Office. National institutes of health: Better oversight needed to help ensure continued progress including women in health research. Washington, DC: US Government Printing Office; 2015. https://www.gao.gov/products/gao-16-13. Accessed 30 Jun 2025.
71. National Institutes of Health. Consideration of sex as a biological variable in NIH-funded research; 2015. https://grants.nih.gov/grants/guide/notice-files/not-od-15-102.html. Accessed 30 Jun 2025.
72. U.S. Food Drug Administration. Enhancing the diversity of clinical trial populations: eligibility criteria, enrollment practices, and trial designs guidance for industry. Food and Drug Administration; 2020.
73. U.S. Food Drug Administration. Diversity action plans to improve enrollment of participants from underrepresented populations in clinical studies draft guidance for industry. Food and Drug Administration; 2024.
74. Health Canada. Inclusion of women in clinical trials during drug development: Government of Canada; 1997. https://www.canada.ca/en/health-canada/services/drugs-health-products/drug-products/applications-submissions/policies/policy-issue-inclusion-women-clinical-tri als-drug-development.html. Accessed 30 Jun 2025.
75. Health Canada. Guidance document: considerations for inclusion of women in clinical trials and analysis of sex differences; 2013. http://www.hc-sc.gc.ca/dhp-mps/prodpharma/applic-demande/guide-ld/clini/womct_femec-eng.php. Accessed 30 Jun 2025.
76. Health Canada. Scientific advisory committee on health products for women—draft terms of reference; 2022. https://www.canada.ca/en/health-canada/services/drugs-health-products/drug-products/scientific-expert-advisory-committees/health-products-women/terms-refere nce.html. Accessed 30 Jun 2025.
77. Greaves L, Brabete AC, Maximos M, Huber E, Li A, Le ML, et al. Sex, gender, and the regulation of prescription drugs: omissions and opportunities. Int J Environ Res Public Health. 2023;20:2962.
78. Health Canada. A guide to the new disaggregated data questionnaire for drug submissions; 2022. https://www.canada.ca/en/health-canada/services/drugs-health-products/drug-pro ducts/applications-submissions/guidance-documents/clinical-trials/disaggregated-data-que stionnaire-submissions.html. Accessed 30 Jun 2025.
79. Ruiz Cantero MT, Angeles PM. European medicines agency policies for clinical trials leave women unprotected. J Epidemiol Community Health. 2006;60:911–3.
80. European Medicines Agency. Gender considerations in the conduct of clinical trials: EMA London; 2005. https://www.ema.europa.eu/en/documents/scientific-guideline/ich-gen der-considerations-conduct-clinical-trials-step-5_en.pdf. Accessed 30 Jun 2025.
81. Consolidated text: Regulation (EU) No 536/2014 of the European Parliament and of the Council of 16 April 2014 on clinical trials on medicinal products for human use, and repealing Directive 2001/20/EC; 2022.

82. European Medicines Agency. Guideline on the investigation of subgroups in confirmatory clinical trials; 2019. https://www.ema.europa.eu/en/documents/scientific-guideline/guideline-investigation-subgroups-confirmatory-clinical-trials_en.pdf. Accessed 30 Jun 2025.
83. Andrade SE, Gurwitz JH, Davis RL, Chan KA, Finkelstein JA, Fortman K, et al. Prescription drug use in pregnancy. Am J Obstet Gynecol. 2004;191:398–407.
84. Ke AB, Greupink R, Abduljalil K. Drug dosing in pregnant women: Challenges and opportunities in using physiologically based pharmacokinetic modeling and simulations. CPT Pharmacometrics Syst Pharmacol. 2018;7:103–10.
85. Jahan N, Went TR, Sultan W, Sapkota A, Khurshid H, Qureshi IA, et al. Untreated depression during pregnancy and its effect on pregnancy outcomes: a systematic review. Cureus. 2021;13: e17251.
86. U.S. Food Drug Administration. Pregnant women: scientific and ethical considerations for inclusion in clinical trials Guidance for Industry; 2018. https://www.fda.gov/regulatory-information/search-fda-guidance-documents/pregnant-women-scientific-and-ethical-considerations-inclusion-clinical-trials. Accessed 30 Jun 2025.
87. National Institutes of Health. Task force on research specific to pregnant women and lactating women; 2018. https://www.nichd.nih.gov/sites/default/files/2018-09/PRGLAC_Report.pdf. Accessed 30 Jun 2025.
88. National Institutes of Health. NIH inclusion outreach toolkit: how to engage, recruit, and retain women in clinical research. https://orwh.od.nih.gov/toolkit. Accessed 30 Jun 2025.
89. World Health Organization. WHO SAGE roadmap for prioritizing uses of COVID-19 vaccines in the context of limited supply: an approach to inform planning and subsequent recommendations based on epidemiological setting and vaccine supply scenarios. Geneva: World Health Organization; 2021.
90. Tannenbaum C, Day D, Matera A. Age and sex in drug development and testing for adults. Pharmacol Res. 2017;121:83–93.
91. Yoon DY, Mansukhani NA, Stubbs VC, Helenowski IB, Woodruff TK, Kibbe MR. Sex bias exists in basic science and translational surgical research. Surgery. 2014;156:508–16.
92. Dekker M, de Vries ST, Versantvoort CHM, Drost-van Velze EGE, Bhatt M, van Meer PJK, et al. Sex proportionality in pre-clinical and clinical trials: an evaluation of 22 marketing authorization application dossiers submitted to the European medicines agency. Front Med (Lausanne). 2021;8: 643028.
93. McCullough LD, de Vries GJ, Miller VM, Becker JB, Sandberg K, McCarthy MM. NIH initiative to balance sex of animals in preclinical studies: generative questions to guide policy, implementation, and metrics. Biol Sex Differ. 2014;5:15.
94. Mauvais-Jarvis F, Arnold AP, Reue K. A guide for the design of pre-clinical studies on sex differences in metabolism. Cell Metab. 2017;25:1216–30.
95. Johnson JL, Greaves L, Repta R. Better science with sex and gender: Facilitating the use of a sex and gender-based analysis in health research. Int J Equity Health. 2009;8:14.
96. Miller LR, Marks C, Becker JB, Hurn PD, Chen WJ, Woodruff T, et al. Considering sex as a biological variable in preclinical research. FASEB J. 2017;31:29–34.
97. U.S. Food Drug Administration. Step 3: clinical research; 2018. https://www.fda.gov/patients/drug-development-process/step-3-clinical-research. Accessed 30 Jun 2025.
98. Lew J, Sanghavi M, Ayers CR, McGuire DK, Omland T, Atzler D, et al. Sex-based differences in cardiometabolic biomarkers. Circulation. 2017;135:544–55.
99. Labots G, Jones A, de Visser SJ, Rissmann R, Burggraaf J. Gender differences in clinical registration trials: is there a real problem? Br J Clin Pharmacol. 2018;84:700–7.
100. World Health Organization. Pharmacovigilance strategies. https://www.who.int/teams/regulation-prequalification/regulation-and-safety/pharmacovigilance/guidance/strategies. Accessed 30 Jun 2025.
101. Härmark L, van Grootheest AC. Pharmacovigilance: methods, recent developments and future perspectives. Eur J Clin Pharmacol. 2008;64:743–52.
102. Kasliwal R. Spontaneous reporting in pharmacovigilance: strengths, weaknesses and recent methods of analysis. J Clin Prev Cardiol. 2012;1:20–3.

103. Lee KMN, Rushovich T, Gompers A, Boulicault M, Worthington S, Lockhart JW, et al. A gender hypothesis of sex disparities in adverse drug events. Soc Sci Med. 2023;339: 116385.
104. Koopmans GT, Lamers LM. Gender and health care utilization: the role of mental distress and help-seeking propensity. Soc Sci Med. 2007;64:1216–30.
105. Xu KT, Borders TF. Gender, health, and physician visits among adults in the United States. Am J Public Health. 2003;93:1076–9.
106. Rushovich T, Gompers A, Lockhart JW, Omidiran I, Worthington S, Richardson SS, et al. Adverse drug events by sex after adjusting for baseline rates of drug use. JAMA Netw Open. 2023;6: e2329074.
107. Racine M, Dion D, Dupuis G, Guerriere DN, Zagorski B, Choiniere M, et al. The Canadian STOP-PAIN project: the burden of chronic pain-does sex really matter? Clin J Pain. 2014;30:443–52.
108. Sattler KM, Deane FP, Tapsell L, Kelly PJ. Gender differences in the relationship of weight-based stigmatisation with motivation to exercise and physical activity in overweight individuals. Health Psychol Open. 2018;5:2055102918759691.
109. Mohammadi F, Kohan S, Mostafavi F, Gholami A. The stigma of reproductive health services utilization by unmarried women. Iran Red Crescent Med J. 2016;18: e24231.
110. Ralli M, Urbano S, Gobbi E, Shkodina N, Mariani S, Morrone A, et al. Health and social inequalities in women living in disadvantaged conditions: a focus on gynecologic and obstetric health and intimate partner violence. Health Equity. 2021;5:408–13.
111. Canadian Institutes of Health Research. Criteria for integration of sex & gender—research with human participants; 2016. https://cihr-irsc.gc.ca/e/49958.html. Accessed 30 Jun 2025.
112. Heidari S, Babor TF, De Castro P, Tort S, Curno M. Sex and gender equity in research: rationale for the SAGER guidelines and recommended use. Res Integr Peer Rev. 2016;1:2.
113. Rich-Edwards JW, Kaiser UB, Chen GL, Manson JE, Goldstein JM. Sex and gender differences research design for basic, clinical, and population studies: Essentials for investigators. Endocr Rev. 2018;39:424–39.
114. Anastario M, Salafia CM, Fitzmaurice G, Goldstein JM. Impact of fetal versus perinatal hypoxia on sex differences in childhood outcomes: developmental timing matters. Soc Psychiatry Psychiatr Epidemiol. 2012;47:455–64.
115. Tobet SA, Handa RJ, Goldstein JM. Sex-dependent pathophysiology as predictors of comorbidity of major depressive disorder and cardiovascular disease. Pflugers Arch. 2013;465:585–94.

Environmental Chemicals and Women's Health from Mid- to Late-Life: Lessons from the Study of Women's Health Across the Nation Multi-Pollutant Study (SWAN-MPS)

Sung Kyun Park

Abstract Midlife has important public health implications, especially for women. Women become more susceptible to cardiometabolic diseases during the menopausal transition due to the hormonal changes that occur in this life stage. Obesity, diabetes, and hypertension in midlife are associated with other age-related diseases such as dementia. Given the key role of midlife as a critical period for early detection and prevention of age-related diseases, there is an urgent need for identifying modifiable risk factors that contribute to disease burden later in life. Midlife may also serve as a window of susceptibility to endogenous environmental chemical exposures because metabolic (e.g., increased bone resorption) and reproductive changes (e.g., menopause) during the menopausal transition increase circulating chemicals concentrations (e.g., lead, per- and polyfluoroalkyl substances (PFAS)). Scientific findings, including those from the Study of Women's Health Across the Nation (SWAN), suggest that environmental chemicals found in personal care products and everyday consumer products play a crucial role in reproductive aging and metabolic health in women. As life expectancy increases and the aging population grows, it is essential to understand modifiable environmental chemicals that influence the timing of menopause and the severity of menopausal symptoms, as these factors can significantly affect women's health later in life. This review introduces findings from the SWAN Multi-Pollutant Study, focusing on PFAS, phthalates, and phenols as exposures and metabolic and reproductive outcomes as health endpoints. This review also discusses lessons learned and individual- and population-level recommendations to reduce the impact of environmental chemicals on women's health.

Keywords Endocrine disrupting chemicals · Environment · Menopause · Midlife · Women's Health

S. K. Park (✉)
Departments of Epidemiology and Environmental Health Sciences, University of Michigan School of Public Health, Ann Arbor, MI, USA
e-mail: sungkyun@umich.edu

H. Lee et al. (eds.), *Sex, Gender, and Emerging Technology in Healthcare: Mitigating Bias and Fostering Equity*, https://doi.org/10.1007/978-981-95-2070-1_8

1 Introduction

Although there is no universally accepted definition of midlife, it is generally considered to span the ages between approximately 40–64 years. Biologically, midlife is characterized by gradual declines in endogenous sex steroid hormones, specifically estrogen in women and testosterone in men [1]. This life stage has a more significant physiological impact on women, as it includes the transition to menopause, which marks the end of menstruation and the clinical termination of reproductive functioning [2]. During the menopausal transition, women become more susceptible to cardiometabolic diseases as hormonal changes lead to alterations in metabolic patterns, including the redistribution of body fat and changes in glucose homeostasis [3–5]. Obesity, diabetes, hypertension, and high low density lipoprotein cholesterol in midlife are associated with other age-related diseases such as dementia [6]. Given midlife's critical role as a period for early detection and prevention of age-related diseases, it is urgent to identify modifiable risk factors that contribute to disease burden later in life. An increasing body of evidence suggests that environmental chemical exposures play a crucial role in reproductive and metabolic health in women. This chapter will provide a brief overview of the role of female reproductive aging during the menopausal transition in increasing the risk of cardiometabolic diseases and endogenous environmental exposures. Additionally, recent scientific findings from the Study of Women's Health Across the Nation (SWAN) addressing the impact of environmental chemical exposures on reproductive and metabolic health in midlife women will be introduced. Finally, the public health implications of the research findings from the SWAN on environmental and women's health will be discussed.

2 Biology of Reproductive Aging in Women

Reproductive aging in women is the physiological aging process in the ovaries characterized by the progressive decline in the quantity and quality of oocytes and ovarian function throughout the reproductive lifespan in women [7, 8]. Reduced ovarian function is evidenced by the occurrence of irregular menstrual cycles, eventually leading to menopause, the complete cessation of menstruation, which is clinically defined retrospectively after 12 consecutive months of amenorrhea [2]. The reduction in ovarian function is characterized by a decrease in follicles, each consisting of a single oocyte surrounded by ovarian granulosa cells, the primary producer of female sex steroid hormone, estradiol (E2) [9]. During the menopause transition, reduced ovarian function leads to alterations in the hypothalamic-pituitary-ovarian axis, resulting in increased secretion of follicle-stimulating hormone (FSH) and a decline in E2 levels [10]. SWAN, a longitudinal study of the natural history of menopause [11], evaluated sex steroid hormones longitudinally, allowing us to understand how E2 an FSH change over the menopausal transition. The rise in FSH typically precedes

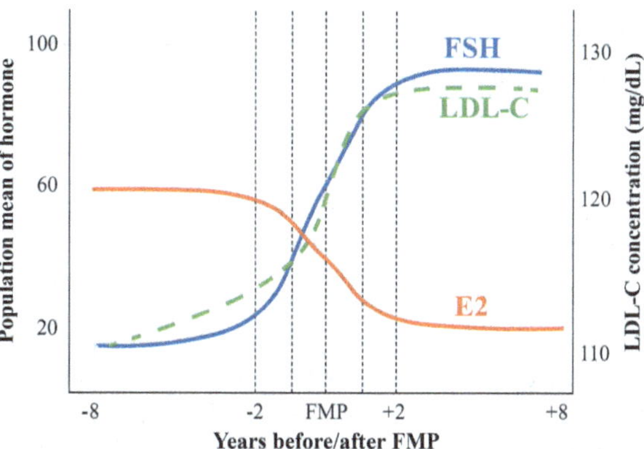

Fig. 1 Schematic of changes in follicle stimulating hormone (FSH), estradiol (E2), and low-density lipoprotein cholesterol (LDL-C) around the final menstrual period (FMP) (modified based on Randolph et al. [12] and Matthews et al. [18])

the decline in E2, with FSH levels beginning to increase about 2 years before the final menstrual period (FMP) [12]. Therefore, decreasing E2 and increasing FSH levels are indicators of key clinical markers of ovarian aging (Fig. 1) [12]. Earlier age of menopause has far-reaching effects on women's health later in life, such as osteoporosis, cardiometabolic diseases, dementia, and premature death [13].

3 Midlife as a Window of Susceptibility to Metabolic Disorders

Reproductive aging and the menopause transition have important public health implications for women. Women become more susceptible to metabolic diseases during the menopausal transition due to the hormonal changes that occur in this life stage [5, 14, 15]. Estrogens play a critical role in maintaining body weight and overall metabolic health [16, 17]. They contribute to the distribution of healthier types of adipose tissue and offer protection against insulin resistance. Moreover, estrogens influence energy balance, lipid metabolism, and glucose homeostasis, thereby reducing the risk of metabolic disorders such as obesity and type 2 diabetes (T2D). Through these mechanisms, estrogens are integral to supporting not only metabolic function but also long-term cardiovascular health and overall well-being. As shown in Fig. 1, FSH starts to increase around 2 years before the FMP and continues to increase at a similar rate until 2 years after the FMP. The pattern of E2 during the menopause transition is opposite. The trajectory of lipids, including low-density lipoprotein cholesterol (LDL-C), parallels those in FSH and E2 [18]. Body weight and body composition (e.g., fat mass) also follow similar patterns during the menopause transition [19]. These longitudinal findings from the SWAN suggest that midlife, more specifically, the menopausal transition, is a window of susceptibility to metabolic disorders especially for women. In turn, the menopause transition is a critical window of opportunity to prevent cardiometabolic disease and other diseases such as osteoporosis and future fractures in women. It is important to identify modifiable risk factors for these health conditions during this critical life stage.

4 Midlife as a Window of Susceptibility to Endogenous Exposures

Midlife may also serve as a window of susceptibility to endogenous environmental toxin exposures. Lead is a prime example. In adults, approximately 95% of the body's lead burden is stored in bone [20]. Thus, bone acts as the major endogenous reservoir for lead. Under normal physiological conditions in healthy adults, lead stores remain in equilibrium. However, the mobilization of lead from bone into the circulation can be accelerated during periods of increased bone turnover, such as pregnancy, lactation, and postmenopause [21]. During the menopause transition, the pattern of bone resorption rate parallels that of FSH, resulting in elevated endogenous lead exposures (Fig. 2). While lead stored in bone may have minimal toxic effects on target organs such as kidneys, heart, and brain, its toxicity can gradually increase during the menopause transition and postmenopause in women.

Per- and polyfluoroalkyl substances (PFAS) is another example. Long-chain PFAS, such as perfluorooctanoic acid (PFOA) and perfluorooctane sulfonic acid (PFOS), are a significant public health concern because they are persistent and do not break down in the environment and in the human body [22]. This is why PFAS are often called as forever chemicals. Approximately over 90% of PFAS are bound to albumin in the blood [23]. Sex differences in PFAS blood concentrations have been reported: women, especially reproductive age women, have lower blood concentrations than men because menstruation is a key elimination pathway of PFAS [24]. PFAS blood concentrations are generally higher in postmenopausal women compared to premenopausal women [25, 26]. As women transition to menopause with cessation of menstruation, this key elimination pathway disappears, resulting in an increase in circulating PFAS concentrations (Fig. 2).

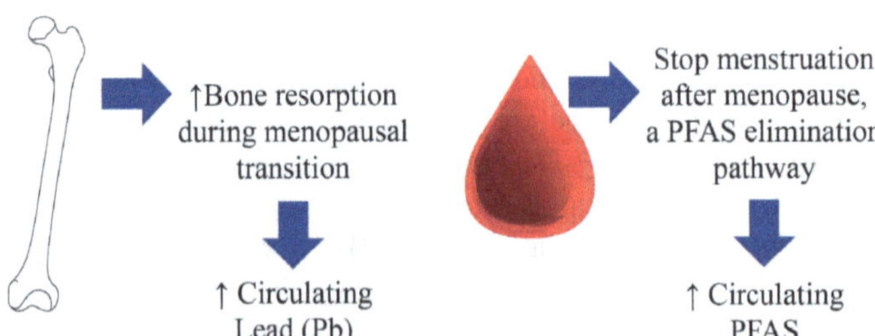

Fig. 2 Pathways of endogenous exposure to lead (Pb) and per- and polyfluoroalkyl substances (PFAS) during the menopausal transition in women

5 Study of Women's Health Across the Nation—Multi-Pollutant Study (SWAN-MPS)

Midlife for women is a period of increased risk for various chronic and age-related diseases, such as metabolic disorders, and heightened exposure to environmental toxicants through endogenous pathways. Despite its significance, midlife remains relatively under-researched in the field of environmental health compared to other life stages.

Several specific questions have not been adequately addressed: *Do exposures to environmental toxicants at midlife cause early menopause, which in turn, increases the risk of cardiometabolic disorders from mid- to late-life?* Addressing these questions is challenging because it requires well-characterized exposures, outcomes, and many other covariates (e.g., confounders, mediators, effect modifiers) collected longitudinally over decades spanning from mid- to late-life. A racially and ethnically diverse, representative sample of premenopausal women is a critical first step. Collection of biospecimens, including blood and urine, is also necessary to assess various environmental toxicants and objective biomarkers for chronic health conditions. The Study of Women's Health Across the Nation—Multi-Pollutant Study (SWAN-MPS) has been established to tackle these challenges.

The SWAN is a comprehensive, multi-site, multi-ethnic longitudinal study aimed at understanding the natural progression of menopause and its impact on subsequent health and age-related chronic disease risk factors (http://www.swanstudy.org) [11] (Fig. 3). The cohort study enrolled 3,302 women between 1996 and 1997 from seven clinic sites, specifically selecting white women and women from one minority group at each site (Black women from Boston, MA, Pittsburgh, PA, Southeast Michigan, MI, and Chicago, IL; Hispanic women from Newark, NJ; Chinese women from Oakland, CA; and Japanese women from Los Angeles, CA). Participants met certain eligibility criteria: they were between 42 and 52 years old, had an intact uterus, experienced at least one menstrual period in the three months prior to the baseline survey, did not take hormone medications (e.g., birth control pills, estrogen, or progesterone) during that time, and identified with the site's designated race/ethnic groups. Data and biological specimens (serum and urine) have been collected almost annually since the study began. Fasting blood and urine were collected, and aliquoted specimens were stored in ultra-low temperature freezers at −80 °C without thawing. All specimens were managed and stored following a standardized protocol. Institutional Review Board approval was secured for each study site, and all participants provided written informed consent at each visit.

In 2016, the SWAN-MPS, funded by the National Institute of Environmental Health Sciences (NIEHS), was launched to investigate the associations between multiple environmental toxicants (including PFAS, polychlorinated biphenyls, organochlorine pesticides, and polybrominated diphenyl ethers in serum; heavy metals, phenols, phthalates, and organophosphate pesticides in urine) and metabolic and reproductive health outcomes in midlife women. Repository samples from the third follow-up visit (Visit 3, 1999–2000) were utilized for environmental exposure

Fig. 3 Screenshot of the SWAN website (left). The map of the SWAN study site (right)

assessment (n = 2,694). Women from Chicago (n = 368) and Newark (n = 278) were excluded due to the lack of urine sample collection at these sites. Additionally, 648 women were excluded due to insufficient serum or urine samples at Visit 3 or insufficient urine samples at Visit 6 (for assessing non-persistent phenols and phthalates), resulting in a sample size of 1,400. Consequently, the SWAN-MPS includes four racial and ethnic groups (White, Black, Chinese, and Japanese) and five study sites (Boston, Pittsburgh, Southeast Michigan, Los Angeles, and Oakland) (Fig. 3). At SWAN-MPS baseline, the mean age was 49.5 years (range: 45–56 years). Approximately half of the women were White, followed by 20% Black, 15% Japanese, and 13% Chinese. They were generally well-educated, with more than 50% holding a college degree or higher. The proportions of past and current smokers were 10% and 27%, respectively. Almost 60% of the women were either overweight or obese.

6 Research Findings in Study of Women's Health Across the Nation—Multi-Pollutant Study (SWAN-MPS)

Here, research findings in SWAN-MPS are summarized: (1) distributions of environmental toxicant concentrations; (2) associations with reproductive aging outcomes; and (3) associations with metabolic outcomes. In this chapter, we restricted to the studies on PFAS, phthalates, and phenols. Studies on other chemicals can be found elsewhere.

6.1 Exposures to Environmental Toxicants

PFAS: As introduce above, PFAS pose a significant public health concern because these chemicals are widely used in everyday products such as food packaging, water-resistant clothing, nonstick cookware, and personal care products (Fig. 4) [22, 27]. PFAS, especially long-chain PFAS (e.g., PFOA and PFOS), are very persistent in

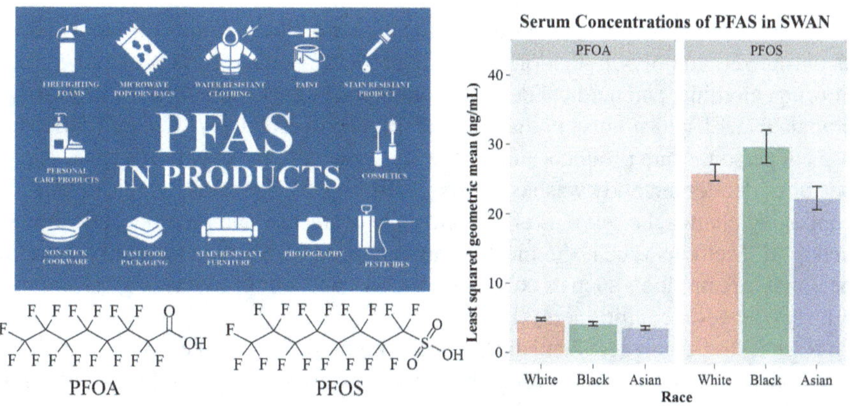

Fig. 4 Sources of PFAS (left upper); chemical structures of two legacy PFAS, PFOA and PFOS (left lower); and serum concentrations of PFOA and PFOS by race in the SWAN-MPS. The PFAS source graph was from https://www.fsawwa.org/page/PFAS

the environment and in the human body because of strong chemical bonds between carbon and fluorine [22, 28]. This property has made these chemicals of significant concern as persistent organic compounds [29]. PFAS have recently received enormous attention in the US because of contaminated drinking water [30] and widespread use that impacts up to 110 million residents in the US (one third of the population) [31]. PFAS have, therefore, been detected in human blood of almost all persons tested in the US [32, 33].

Women may be exposed to PFAS more due to their higher usage of personal care products, including shampoo, nail polish, and makeup [34]. A total of 11 PFAS compounds were measured using serum samples collected in 1999–2000 at SWAN-MPS baseline. This chapter focuses on two legacy compounds, PFOA and PFOS, as these are prevalent in exposure levels and most extensively studies in previous studies. The SWAN study is uniquely positioned to provide information on racial/ethnic and geographic differences in PFAS exposure. Race/ethnicity was identified as a major determinant of PFAS blood concentrations: White women had higher concentrations of PFOA, whereas Black women had higher concentrations of PFOS [35] (Fig. 4). In particular, after adjustment for age, site, race, education, and body mass index (BMI), the least squared geometric mean concentration of PFOS was 29.7 ng/mL (95% confidence interval [CI], 27.4–32.1) in Black women, whereas it was 25.8 ng/mL (95% CI, 24.8–27.1) in White women, and 22.2 ng/mL (95% CI, 20.5–24.0) in Asian women (Fig. 4). This suggests that the sources of these two compounds are different and that the patterns of exposure are race-specific [36]. The study also confirmed that menstrual bleeding and childbirth are associated with lower concentrations of PFAS, indicating that these reproductive processes in women are important elimination pathways for PFAS [24, 37].

Phthalates and Phenols: Various everyday products also contain phthalates (di-esters of 1,2-benzenedicarboxylic acid) and phenols. For example, di(2-ethylhexyl) phthalate (DEHP) and other high-molecular weight phthalates are commonly added as plasticizers to polyvinyl chloride plastic products, such as food packaging, vinyl flooring, clothing, and medical devices (e.g., blood bags) [38]. Furthermore, di-ethyl phthalate (DEP), di-n-butyl phthalate (DnBP), and di-isobutyl phthalate (DiBP) are used in personal care products and cosmetics, especially those with fragrances (e.g., perfume, deodorant, body washes, lotions) [39] (Fig. 5). Parabens, which are phenol-containing chemicals, serve as preservatives in cosmetics and pharmaceutical products [40]. Methylparaben, the most commonly used paraben, is primarily found in personal care products such as cosmetics, lotions, and hair products (e.g., shampoos, relaxers, leave-in conditioners) (Fig. 5).

Recent studies suggest that Black women are more likely to be exposed to phthalates, phenols, and parabens through hair product use compared to other racial/ethnic groups [41–43]. In the SWAN-MPS studies, exposures to phthalates and phenols were assessed by measuring urinary metabolites in samples collected in 1999–2000. Women who were younger, current smokers, or obese generally had higher concentrations of phthalates. As anticipated, Black women showed higher concentrations of urinary metabolites of phthalates, particularly mono-ethyl phthalate (MEP), a metabolite of DEP [44, 45]. After adjusting for age, site, race/ethnicity, education, and BMI, the least squared geometric mean concentration of MEP was 240 ng/mL (95% CI, 198–287) in Black women, compared to 97 ng/mL (95% CI, 86–109) in White women and 49 ng/mL (95% CI, 40–59) in Asian women (Fig. 5). A similar pattern was observed for urinary methylparaben concentrations [46]: 290 ng/mL

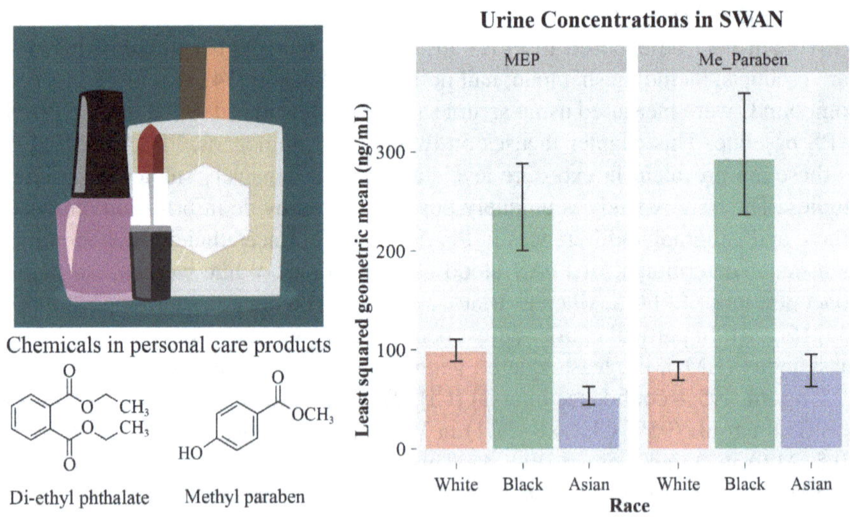

Fig. 5 Personal care products (e.g., cosmetics, lotions) as sources of phthalates and phenols (left upper); chemical structures of di-ethyl phthalate (DEP) and methylparaben (Me-paraben) (left lower); and serum concentrations of mono-ethyl phthalate (MEP, a metabolite of DEP) and Me-paraben by race in the SWAN-MPS. The phthalates and phenols source graph was obtained from Pixabay (www.pixabay.com) and is used under a free-for-use license

(95% CI, 235–358) in Black women, 74 ng/mL (95% CI, 66–85) in White women, and 74 ng/mL (95% CI, 60–92) in Asian women (Fig. 5). These findings indicate the need for greater efforts to reduce not only sex differences but also racial/ethnic disparities in exposure to personal care product chemicals within women through stricter regulatory standards for personal care products.

6.2 Exposure to Environmental Chemicals and Reproductive Aging

Reproductive aging refers to the gradual decline in ovarian function with age [7]. It is characterized by reduced follicle numbers and diminished oocyte quality [47]. Recent epidemiologic investigations from the major cohort studies of the menopausal transition including the SWAN strengthen the evidence that the timing of reproductive aging influence women's risk of chronic disease and their long-term health with aging [48–50]. Most previous epidemiologic studies examining the associations between environmental chemicals and reproductive aging outcomes have been cross-sectional with retrospective recall of menopausal parameters including age at menopause, raising concerns related to causal inference. The SWAN assessed the timing of menopause through the bleeding questions on the annual visit questionnaire, yielding prospective classification of menopausal status and the date of the FMP. The SWAN annual visits also assessed sex hormones (e.g., E2, FSH, testosterone) and anti-Müllerian hormone (AMH) as reproductive outcomes. AMH is a dimeric glycoprotein exclusively produced by granulosa cells of preantral and small antral follicles [51, 52]. AMH serum levels gradually decrease with the decrease in the number of antral follicles with age and become undetectable near menopause [53], and thus are considered a novel biomarker for reproductive aging. Table 1 summarizes the SWAN-MPS publications on reproductive aging outcomes.

We examined timing of menopause as incident natural menopause using time-to-event survival analysis. For PFAS, both PFOS and PFOA were significantly associated with higher risk of incident natural menopause [54], suggesting that these major legacy PFAS may lead to earlier natural menopause, a risk factor for adverse health outcomes including bone loss in later life. As exposure to individual PFAS compounds is not independent, we evaluate all PFAS compounds as mixtures and identified four PFAS exposure clusters using *k*-means (low, moderate low, moderate high, high). Compared with the low cluster, the high cluster had a hazard ratio (HR) for natural menopause of 1.63 (95% CI, 1.08–2.45), which is equivalent to 2.0 years earlier median time to natural menopause. This estimate is even larger than the effect of cigarette smoking (1.1 years comparing current smokers vs. never smokers in our sample) [54]. In a subsequent study, we found that generally, PFAS were positively associated with FSH and inversely associated with E2 [55], confirming that PFAS affect reproductive aging.

Table 1 Summary of SWAN-MPS studies examining reproductive aging outcomes

Reference (first author, year)	Exposures	Outcomes (effect estimates)	Major findings
Ding, 2020 [54]	PFAS	Timing of menopause (HR tertile 3 vs. tertile 1)	PFOA: **1.31 (1.04 to 1.65)** PFOS: **1.26 (1.02 to 1.57)** PFAS mixture: **1.63 (1.08 to 2.45)**
Harlow, 2021 [55]	PFAS	Sex hormones (Percent difference per doubling of exposure)	Estradiol (E2): PFOA: −2.43% (−4.97 to 0.18) PFOS: −1.72% (−4.20 to 0.83) FSH: PFOA: **3.12% (0.37 to 5.95)** PFOS: **3.03% (0.37 to 5.76)**
Ding, 2023 [56]	Phthalates	Timing of menopause (HR per doubling of exposure)	ΣDEHP: 0.99 (0.94 to 1.05) ΣLMW: 0.98 (0.93 to 1.03) ΣHMW: 0.98 (0.94 to 1.06)
		Hormones (Percent difference per doubling of exposure)	Testosterone: MCOP: **−2.08% (−3.66 to −0.47)** MnBP: **−1.99% (−3.66 to −0.47)** ΣDEHP: −0.50% (−1.89 to 0.92) ΣLMW: −0.91% (−2.3 to 0.50) ΣHMW: −0.52% (−2.06 to 1.06) Anti-Müllerian hormone (AMH): MECPP: **−14.3% (−24.1 to −3.14)** MEHHP: **−15.6% (−24.6 to −5.50)** MEOHP: **−13.5% (−22.9 to −2.90)** ΣDEHP: **−15.0% (−24.6 to −4.26)**

Numbers in bold indicate statistically significant findings at $p < 0.05$

PFAS Per- and polyfluoroalkyl substances, *PFOA* perfluorooctanoic acid, *PFOS* perfluorooctane sulfonic acid, *HR* hazard ratio, *FSH* follicle-stimulating hormone, *DEHP* di(2-ethylhexyl) phthalate, *LMW* low-molecular-weight, *HMW* high-molecular-weight, *MCOP* mono-carboxyoctyl phthalate, *MnBP* mono-n-butyl phthalate, *MECPP* mono-(2-ethyl-5-carboxylpentyl) phthalate, *MEHHP* mono-(2-ethyl-5-hydroxylhexyl) phthalate, *MEOHP* mono-(2-ethyl-5-oxohexyl) phthalate

For phthalates, several urinary metabolites were significantly associated with lower levels of testosterone: Each doubling of urinary concentrations of mono-carboxyoctyl phthalate (MCOP) and mono-n-butyl phthalate (MnBP) was associated with 2.08% lower (95% CI, −3.66 to −0.47) and 1.99% lower (−3.66 to −0.47) testosterone levels [56]. These findings are consistent with the notion of antiandrogenic properties of phthalates. Additionally, we observed inverse associations between DEHP metabolites and AMH: percent differences in AMH per doubling of exposure were −14.26% (95% CI, −24.10 to −3.14) for mono-(2-ethyl-5-carboxylpentyl) phthalate (MECPP), −15.58% (95% CI, −24.59 to −5.50) for mono-(2-ethyl-5-hydroxylhexyl) phthalate (MEHHP), and −13.50% (95% CI, −22.93 to −2.90) for mono-(2-ethyl-5-oxohexyl) phthalate (MEOHP), respectively. This is the first evidence showing the association between phthalate exposure and diminished ovarian reserve in midlife women. However, there was no statistically significant association between phthalates and timing of menopause.

7 Exposure to Environmental Chemicals and Metabolic Outcomes

Metabolic diseases including T2D, obesity and metabolic syndrome (MetS) have been increasing dramatically over the past several decades in the United States (US) and in the world. Currently 37.1 million adults in the US are diagnosed with T2D (14.7% of the population aged 18 years and older) with the prevalence being 29.2% in people aged 65 years or older [57]. According to the World Health Organization report in 2022, over 40% of adults were overweight (BMI 25–29.9 kg/m^2) and 16% were obese (BMI $\geq$ 30 kg/m^2) [58]. Obesity is a more serious public health concern in developed countries. In the US, the prevalence of obesity in adults during 2021–2023 was approximately 40% and middle-aged adults ages 40–59 years had the highest prevalence of obesity (46.4%) than other age groups [59]. Women are more susceptible to severe obesity (BMI $\geq$ 40 kg/m^2) with the prevalence of 12.1% compared to men (6.7%) [59]. In the US, approximately 35% of adults have MetS with over half of adults aged 60 years and older having this condition [60]. MetS increases the risk of developing cardiovascular disease (CVD) and T2D [61]. It is well understood that T2D and components of MetS are caused by a combination of genetic and environmental factors. Although genetic factors play an important role, they have largely not changed over the past several decades, while the prevalence of T2D and MetS has skyrocketed [62, 63], Known genes account for only a small proportion of the variance in incidence of T2D and components of the MetS. Therefore, the current epidemic of these metabolic diseases is more likely due to major changes in environmental factors. Increased energy intake and a decrease in physical activity levels are important contributors to the rise [64]. The widespread exposure to environmental chemicals may also be a key contributor to the growing epidemics of T2D and MetS [65, 66]. A particular class of endocrine disrupting chemicals that have the ability to promote metabolic alterations causing obesity, T2D and MetS are called metabolism-disrupting chemicals (MDCs) [65, 67].

Similar to the reproductive aging outcomes, we evaluated metabolic outcomes longitudinally in the SWAN-MPS in relation to various MDCs. Table 2 summarizes the SWAN-MPS publications on metabolic outcomes.

For PFAS, we have found compelling increased risks of incident T2D with several PFAS including PFOA and PFOS [68], suggesting that these MDCs may lead to earlier development of T2D. Exposure to PFAS mixtures was associated with a HR of 2.62 (95% CI, 1.12–6.20) comparing the top vs the bottom tertiles for all PFAS using quantile g-computation, suggesting a potential additive or synergistic effect [68]. Similarly, several PFAS compounds were significantly associated with incident hypertension. In the PFAS mixtures analysis, the highest tertile for all PFAS compounds comparing the lowest tertile was associated with a HR of 1.71 (95% CI, 1.15–2.54) [69]. PFAS were also associated with greater body size and body fat as well as accelerated increase in obesity measures over time [70]. At baseline, women in the highest tertile of PFAS were heavier than those in the lowest tertile: the adjusted geometric mean (GM) weights were 73.9 versus 69.6 kg (ratio of GM 1.06; 95% CI, 1.03–1.09) for PFOS and 74.0 versus 69.4 kg (ratio of GM 1.07; 95% CI, 1.03–1.10) for PFOA. Women in the highest tertile had an annual weight increase rate of 0.33% (95% CI, 0.27–0.40%) over 9 years of follow-up, compared to 0.10% (95% CI, 0.04–0.17%) in the lowest tertile for PFOS; and 0.21% (95% CI, 0.15–0.28%) compared to 0.13% (95% CI, 0.07–0.20%) for PFOA. Similar patterns were observed for waist circumference and fat mass. Furthermore, we found positive associations between PFAS and trajectories of total and LDL-C, suggesting that PFAS may disturb lipid homeostasis [71]. For the join effect of PFAS mixtures, women in the high exposure group comparing with the low exposure group had odds ratios (ORs) of 1.69 (95% CI, 1.36–2.12) for total cholesterol and 1.79 (95% CI, 1.44–2.22) for LDL-C, respectively.

For phthalates, we observed that some phthalates were associated with a higher risk of T2D development, especially in White women [44]. Each doubling of urinary concentrations of mono-isobutyl phthalate (MiBP), mono(3-carboxypropyl) phthalate (MCPP), and the sum of high-molecular weight phthalates was associated with HRs of 1.63 (1.18–2.25), 1.50 (1.06–2.12), and 1.77 (1.27–2.46), respectively, in White women. These associations were not observed in Black or Asian women. It is unclear why White women had larger effects. One plausible explanation is that the study commenced monitoring women during midlife, despite phthalate exposure beginning early in life and persisting over time. Due to this design choice, the analysis could not include women who had been exposed to high levels of phthalates and subsequently developed T2D at an earlier age. Consequently, these women, who may have been at an increased risk of T2D associated with phthalate exposure, were not represented in this study population. Furthermore, these unrepresented women might have been disproportionately Black, as Black women tend to develop T2D at a younger age and are exposed to higher levels of various phthalates compared to White women [44].

Table 2 Summary of SWAN-MPS studies examining metabolic outcomes

Reference (first author, year)	Exposures	Outcomes (effect estimates)	Major findings
Park, 2022 [68]	PFAS	Incident diabetes (HR, T3 vs. T1)	PFOA: **1.67 (1.21 to 2.31)** PFOS: 1.25 (0.90 to 1.74) PFAS mixture: **2.62 (1.12 to 6.20)**
Ding, 2022 [69]	PFAS	Incident hypertension (HR, T3 vs. T1)	PFOA: **1.47 (1.24 to 1.75)** PFOS: **1.42 (1.19 to 1.68)** PFAS mixture: **1.71 (1.15 to 2.54)**
Ding, 2021 [70]	PFAS	Weight (ratio of GM, T3 vs. T1 in baseline; percent change for the slope of annual rate)	Baseline: PFOA: **1.07 (1.03 to 1.10)** PFOS: **1.06 (1.03 to 1.09)** Annual rate of change over 9 y: PFOA: **0.21% (T3) versus 0.13% (T1)** PFOS: **0.33% (T3) versus 0.10% (T1)**
		Waist circumference (ratio of GM, T3 vs. T1 in baseline; percent change for the slope of annual rate)	Baseline: PFOA: **1.04 (1.02 to 1.07)** PFOS: **1.04 (1.02 to 1.07)** Annual rate of change over 9 y: PFOA: 0.38% (T3) versus 0.34% (T1) PFOS: 0.45% (T3) versus 0.33% (T1)

(continued)

Table 2 (continued)

Reference (first author, year)	Exposures	Outcomes (effect estimates)	Major findings
		Fat mass (ratio of GM, T3 vs. T1 in baseline; percent change for the slope of annual rate)	PFOA: **1.12 (1.06 to 1.17)** PFOS: **1.10 (1.05 to 1.16)** Annual rate of change over 14.5 y: PFOA: **0.21% (T3) versus 0.14% (T1)** PFOS: **0.31% (T3) versus 0.06% (T1)**
Kang, 2023 [71]	PFAS	Total cholesterol (OR comparing high vs. low trajectories, T3 vs. T1)	PFOA: 0.90 (0.70 to 1.15) PFOS: **1.70 (1.33 to 2.17)** PFAS mixture: **1.69 (1.36 to 2.12)**
		LDL cholesterol (OR comparing high vs. low trajectories, T3 vs. T1)	PFOA: 1.05 (0.83 to 1.35) PFOS: **1.94 (1.54 to 2.45)** PFAS mixture: **1.79 (1.44 to 2.22)**

(continued)

Table 2 (continued)

Reference (first author, year)	Exposures	Outcomes (effect estimates)	Major findings
Peng, 2023 [44]	Phthalates	Incident diabetes (HR per doubling of exposure)	All women: MiBP: 1.19 (0.94 to 1.49) MCPP: 1.15 (0.91 to 1.51) ΣDEHP: 1.00 (0.84 to 1.20) ΣLMW: 1.04 (0.88 to 1.24) ΣHMW: 1.19 (0.94 to 1.51) White women: MiBP: **1.63 (1.18 to 2.25)** MCPP: **1.50 (1.06 to 2.12)** ΣDEHP: 1.14 (0.88 to 1.46) ΣLMW: 1.15 (0.89 to 1.47) ΣHMW: **1.77 (1.27 to 2.46)**
Lee, 2021 [72]	Phenols	Incident diabetes (HR, T3 vs. T1)	BPA: 1.31 (0.94 to 1.83) BPF: 1.01 (0.74 to 1.83) Triclosan: 1.14 (0.84 to 1.56) Me-paraben: **0.40 (0.29 to 0.56)** Pr-paraben: **0.42 (0.30 to 0.58)** BP3: **0.55 (0.39 to 0.80)**

Numbers in bold indicate statistically significant findings at $p < 0.05$

PFAS Per- and polyfluoroalkyl substances, *PFOA* perfluorooctanoic acid, *PFOS* perfluorooctane sulfonic acid, *HR* hazard ratio, *T3* tertile 3, *T1* tertile 1, *GM* geometric mean, *LDL* low-density lipoprotein, *OR* odd ratio, *MiBP* mono-isobutyl phthalate, *MCPP* mono(3-carboxypropyl) phthalate, *DEHP* di(2-ethylhexyl) phthalate, *LMW* low-molecular-weight, *HMW* high-molecular-weight, *BPA* bisphenol A, *BPF* bisphenol F

For phenols, we did not observe positive associations with incident T2D, suggesting no diabetogenic effects of these chemicals commonly found in everyday products [72]. Interestingly, some phenols showed inverse associations. Women in the highest tertiles comparing with the lowest tertiles had HRs of 0.40 (95% CI, 0.29–0.56) for methylparaben, 0.42 (95% CI, 0.30–0.58) for propylparaben, and 0.55 (95% CI, 0.39–0.80) for benzophenone-3. These inverse associations may be biologically plausible as these chemicals have antiandrogenic and estrogenic properties [73] and can activate the expression of peroxisome proliferator-activated receptor-γ (PPAR-γ), which could prevent insulin resistance and development of T2D [74].

8 Conclusions

Scientific findings, including those from the SWAN-MPS, suggest that environmental chemicals commonly found in personal care products and everyday consumer products play a crucial role in reproductive aging and metabolic health in midlife women. PFAS, phthalates, and phenols are among the chemical classes often detected in such products. Numerous other potentially toxic chemicals are intentionally added to extend shelf life, improve or mask scents, enhance convenience, and increase product durability. Women are at a higher risk for exposure to these chemicals as they typically use more personal care products; on average, an adult woman uses 12 different personal care products containing over 168 unique chemicals daily [75]. If exposure to these environmental chemicals can impact reproductive aging during the menopause transition and subsequently increase the risk of developing cardiometabolic and other chronic diseases, what can women do to counteract this?

Reducing exposure to environmental chemicals at the individual-level is challenging but remains an effective strategy for preventing chronic diseases through behavioral changes. Diet and the use of everyday products are major sources of PFAS exposure, which is often associated with 'convenience'. For instance, food wrappers, non-grease papers, and other food contact materials contain PFAS. Additionally, waterproof and stain-proof fabrics, as well as some cosmetics, also contain these chemicals. While it is impossible to completely avoid exposure, it is crucial to recognize these sources and adopt a mindset that occasionally values 'taking inconvenience' over convenience. Active consumer campaigns can also play a vital role in pushing industry towards change. When consumers raise their collective voice, industries often respond by voluntarily eliminating toxic chemicals from their products. A recent example includes major restaurant chains and retail stores committing to eliminating PFAS in their packaging [76–78].

However, a more impactful strategy involves changing policies to limit PFAS and other toxic chemicals in the air, drinking water, foods, and other environmental sources. Once released into the environment, the cost of eliminating PFAS becomes increasing uncontrollable [79]. Determining who is responsible for these societal cost remains a complex issue. Therefore, it is essential to build public consensus and seek effective solutions to reduce these costs while protecting both human health and the environment. A recent study has shown that stricter chemical regulations can reduce the overall exposure of the population to toxic chemicals. For example, California's Proposition 65 (Prop 65), which was designed to safeguard residents from chemicals that could cause cancer or affect birth and reproductive outcomes, has led to a decrease in exposure levels to toxic chemicals including phthalates and phenols, not only among Californians but also nationally [80]. Women, in particular, may benefit from these changes the most.

Menopause is not a disease, and consequently, it has received less attention as a significant lifestage. However, as life expectancy increases and the aging population grows, it is essential to understand the modifiable risk factors that influence the timing of menopause and the severity of menopausal symptoms, as these factors can significantly affect women's health later in life. Environmental chemicals play a crucial role in this process.

Acknowledgments The Study of Women's Health Across the Nation Multi-Pollutant Study (SWAN-MPS) has grant support from the National Institutes of Health (NIH), DHHS, through the National Institute on Aging (NIA), the National Institute of Nursing Research (NINR), the NIH Office of Research on Women's Health (ORWH), and the National Institute of Environmental Health Sciences (NIEHS) (Grants U01NR004061; U01AG012505, U01AG012535, U01AG012531, U01AG012539, U01AG012546, U01AG012553, U01AG012554, U01AG012495, U19AG063720, U01AG017719, R01-ES026578, R01-ES026964, R01 ES035087). The content of this book chapter is solely the responsibility of the authors and does not necessarily represent the official views of the NIA, NINR, ORWH, NIEHS or the NIH.

References

1. Infurna FJ, Gerstorf D, Lachman ME. Midlife in the 2020s: opportunities and challenges. Am Psychol. 2020;75:470–85.
2. Harlow SD, Paramsothy P. Menstruation and the menopausal transition. Obstet Gynecol Clin North Am. 2011;38:595–607.
3. El Khoudary SR, Aggarwal B, Beckie TM, Hodis HN, Johnson AE, Langer RD, et al. Menopause transition and cardiovascular disease risk: implications for timing of early prevention: a scientific statement from the American Heart Association. Circulation. 2020;142:e506–32.
4. El Khoudary SR, Nasr A. Cardiovascular disease in women: does menopause matter? Curr Opin Endocr Metab Res. 2022;27:100419.
5. Polotsky HN, Polotsky AJ. Metabolic implications of menopause. Semin Reprod Med. 2010;28:426–34.
6. Livingston G, Huntley J, Liu KY, Costafreda SG, Selbæk G, Alladi S, et al. Dementia prevention, intervention, and care: 2024 report of the Lancet standing Commission. Lancet. 2024;404:572–628.
7. Broekmans FJ, Soules MR, Fauser BC. Ovarian aging: mechanisms and clinical consequences. Endocr Rev. 2009;30:465–93.
8. Wang X, Wang L, Xiang W. Mechanisms of ovarian aging in women: a review. J Ovarian Res. 2023;16:67.
9. Hall JE. Endocrinology of the menopause. Endocrinol Metab Clin North Am. 2015;44:485–96.
10. Tanbo TG, Fedorcsak PZ. Can time to menopause be predicted? Acta Obstet Gynecol Scand. 2021;100:1961–8.

11. Sowers MR, Crawford SL, Sternfeld B, Morganstein D, Gold EB, Greendale GA. SWAN: A multicenter, multiethnic, community-based cohort study of women and the menopausal transition. In: Lobo RA, Kelsey J, Marcus R, editors. Menopause: biology and pathobiology. San Diego, CA: Academic Press; 2000. p. 175–88.
12. Randolph JF Jr, Zheng H, Sowers MR, Crandall C, Crawford S, Gold EB, et al. Change in follicle-stimulating hormone and estradiol across the menopausal transition: effect of age at the final menstrual period. J Clin Endocrinol Metab. 2011;96:746–54.
13. Gold EB, Crawford SL, Avis NE, Crandall CJ, Matthews KA, Waetjen LE, et al. Factors related to age at natural menopause: longitudinal analyses from SWAN. Am J Epidemiol. 2013;178:70–83.
14. Davis SR, Castelo-Branco C, Chedraui P, Lumsden MA, Nappi RE, Shah D, et al. Understanding weight gain at menopause. Climacteric. 2012;15:419–29.
15. Wildman RP, Sowers MR. Adiposity and the menopausal transition. Obstet Gynecol Clin North Am. 2011;38:441–54.
16. Bracht JR, Vieira-Potter VJ, De Souza SR, Öz OK, Palmer BF, Clegg DJ. The role of estrogens in the adipose tissue milieu. Ann N Y Acad Sci. 2020;1461:127–43.
17. Zhu J, Zhou Y, Jin B, Shu J. Role of estrogen in the regulation of central and peripheral energy homeostasis: from a menopausal perspective. Ther Adv Endocrinol Metab. 2023;14:20420188231199360.
18. Matthews KA, Crawford SL, Chae CU, Everson-Rose SA, Sowers MF, Sternfeld B, et al. Are changes in cardiovascular disease risk factors in midlife women due to chronological aging or to the menopausal transition? J Am Coll Cardiol. 2009;54:2366–73.
19. Greendale GA, Sternfeld B, Huang M, Han W, Karvonen-Gutierrez C, Ruppert K, et al. Changes in body composition and weight during the menopause transition. JCI Insight. 2019;4:e124865.
20. Saltzman BE, Gross SB, Yeager DW, Meiners BG, Gartside PS. Total body burdens and tissue concentrations of lead, cadmium, copper, zinc, and ash in 55 human cadavers. Environ Res. 1990;52:126–45.
21. Hu H, Rabinowitz M, Smith D. Bone lead as a biological marker in epidemiologic studies of chronic toxicity: conceptual paradigms. Environ Health Perspect. 1998;106:1–8.
22. Agency for Toxic Substances and Disease Registry (ATSDR). Toxicological profile for Perfluoroalkyls. Atlanta, GA: U.S. Department of Health and Human Services, Public Health Service; 2021.
23. Han X, Snow TA, Kemper RA, Jepson GW. Binding of perfluorooctanoic acid to rat and human plasma proteins. Chem Res Toxicol. 2003;16:775–81.
24. Wong F, MacLeod M, Mueller JF, Cousins IT. Enhanced elimination of perfluorooctane sulfonic acid by menstruating women: evidence from population-based pharmacokinetic modeling. Environ Sci Technol. 2014;48:8807–14.
25. Knox SS, Jackson T, Javins B, Frisbee SJ, Shankar A, Ducatman AM. Implications of early menopause in women exposed to perfluorocarbons. J Clin Endocrinol Metab. 2011;96:1747–53.
26. Taylor KW, Hoffman K, Thayer KA, Daniels JL. Polyfluoroalkyl chemicals and menopause among women 20–65 years of age (NHANES). Environ Health Perspect. 2014;122:145–50.
27. The Interstate Technology and Regulatory Council (ITRC). History and use of per-and polyfluoroalkyl substances (PFAS); 2020.
28. Buck RC, Franklin J, Berger U, Conder JM, Cousins IT, de Voogt P, et al. Perfluoroalkyl and polyfluoroalkyl substances in the environment: terminology, classification, and origins. Integr Environ Assess Manag. 2011;7:513–41.
29. Post GB, Gleason JA, Cooper KR. Key scientific issues in developing drinking water guidelines for perfluoroalkyl acids: contaminants of emerging concern. PLoS Biol. 2017;15:e2002855.
30. Hu XC, Andrews DQ, Lindstrom AB, Bruton TA, Schaider LA, Grandjean P, et al. Detection of poly- and perfluoroalkyl substances (PFASs) in U.S. drinking water linked to industrial sites, military fire training areas, and wastewater treatment plants. Environ Sci Technol Lett. 2016;3:344–50.
31. Evans S, Andrews D, Stoiber T, Naidenko O. PFAS contamination of drinking water far more prevalent than previously reported. Environmental Working Group; 2020. https://www.ewg.org/research/national-pfas-testing. Accessed 30 Jun 2025.

32. National Center for Environmental Health (U.S.). Division of Laboratory Sciences. Fourth national report on human exposure to environmental chemicals: updated tables, vol. 1; 2019.
33. National Center for Environmental Health (U.S.). Division of laboratory sciences, national health and nutrition examination survey (U.S.). Fourth national report on human exposure to environmental chemicals. Atlanta, GA; 2009.
34. Hall AM, Ashley-Martin J, Lei Liang C, Papandonatos GD, Arbuckle TE, Borghese MM, et al. Personal care product use and per- and polyfluoroalkyl substances in pregnant and lactating people in the maternal-infant research on environmental chemicals study. Environ Int. 2024;193:109094.
35. Park SK, Peng Q, Ding N, Mukherjee B, Harlow SD. Determinants of per- and polyfluoroalkyl substances (PFAS) in midlife women: evidence of racial/ethnic and geographic differences in PFAS exposure. Environ Res. 2019;175:186–99.
36. Trudel D, Horowitz L, Wormuth M, Scheringer M, Cousins IT, Hungerbühler K. Estimating consumer exposure to PFOS and PFOA. Risk Anal. 2008;28:251–69.
37. Berg V, Nøst TH, Huber S, Rylander C, Hansen S, Veyhe AS, et al. Maternal serum concentrations of per- and polyfluoroalkyl substances and their predictors in years with reduced production and use. Environ Int. 2014;69:58–66.
38. National Research Council (US) Committee on the Health Risks of Phthalates. Phthalates and cumulative risk assessment: the tasks ahead. Washington (DC): National Academies Press (US); 2008.
39. Zota AR, Calafat AM, Woodruff TJ. Temporal trends in phthalate exposures: findings from the national health and nutrition examination survey, 2001–2010. Environ Health Perspect. 2014;122:235–41.
40. Wei F, Mortimer M, Cheng H, Sang N, Guo LH. Parabens as chemicals of emerging concern in the environment and humans: a review. Sci Total Environ. 2021;778:146150.
41. Helm JS, Nishioka M, Brody JG, Rudel RA, Dodson RE. Measurement of endocrine disrupting and asthma-associated chemicals in hair products used by black women. Environ Res. 2018;165:448–58.
42. James-Todd T, Senie R, Terry MB. Racial/ethnic differences in hormonally-active hair product use: a plausible risk factor for health disparities. J Immigr Minor Health. 2012;14:506–11.
43. James-Todd T, Connolly L, Preston EV, Quinn MR, Plotan M, Xie Y, et al. Hormonal activity in commonly used Black hair care products: evaluating hormone disruption as a plausible contribution to health disparities. J Expo Sci Environ Epidemiol. 2021;31:476–86.
44. Peng MQ, Karvonen-Gutierrez CA, Herman WH, Mukherjee B, Park SK. Phthalates and incident diabetes in midlife women: The Study of Women's Health Across the Nation (SWAN). J Clin Endocrinol Metab. 2023;108:1947–57.
45. Peng MQ, Karvonen-Gutierrez CA, Herman WH, Mukherjee B, Park SK. Phthalate exposure is associated with more rapid body fat gain in midlife women: The Study of Women's Health Across the Nation (SWAN) multi-pollutant study. Environ Res. 2023;216:114685.
46. Lee S, Karvonen-Gutierrez C, Mukherjee B, Herman WH, Park SK. Race-specific associations of urinary phenols and parabens with adipokines in midlife women: The Study of Women's Health Across the Nation (SWAN). Environ Pollut. 2022;303:119164.
47. Djahanbakhch O, Ezzati M, Zosmer A. Reproductive ageing in women. J Pathol. 2007;211:219–31.
48. Bleil ME, Gregorich SE, McConnell D, Rosen MP, Cedars MI. Does accelerated reproductive aging underlie premenopausal risk for cardiovascular disease? Menopause. 2013;20:1139–46.
49. Greendale GA, Huang MH, Wight RG, Seeman T, Luetters C, Avis NE, et al. Effects of the menopause transition and hormone use on cognitive performance in midlife women. Neurology. 2009;72:1850–7.
50. Janssen I, Powell LH, Crawford S, Lasley B, Sutton-Tyrrell K. Menopause and the metabolic syndrome: The Study of Women's Health Across the Nation. Arch Intern Med. 2008;168:1568–75.
51. Moolhuijsen LME, Visser JA. Anti-Müllerian hormone and ovarian reserve: Update on assessing ovarian function. J Clin Endocrinol Metab. 2020;105:3361–73.

52. Visser JA, Schipper I, Laven JS, Themmen AP. Anti-Müllerian hormone: an ovarian reserve marker in primary ovarian insufficiency. Nat Rev Endocrinol. 2012;8:331–41.
53. Sowers MR, Eyvazzadeh AD, McConnell D, Yosef M, Jannausch ML, Zhang D, et al. Anti-Mullerian hormone and inhibin B in the definition of ovarian aging and the menopause transition. J Clin Endocrinol Metab. 2008;93:3478–83.
54. Ding N, Harlow SD, Randolph JF, Calafat AM, Mukherjee B, Batterman S, et al. Associations of perfluoroalkyl substances with incident natural menopause: The Study of Women's Health Across the Nation. J Clin Endocrinol Metab. 2020;105:e3169–82.
55. Harlow SD, Hood MM, Ding N, Mukherjee B, Calafat AM, Randolph JF, et al. Per- and polyfluoroalkyl substances and hormone levels during the menopausal transition. J Clin Endocrinol Metab. 2021;106:e4427–37.
56. Ding N, Zheutlin E, Harlow SD, Randolph JF, Jr., Mukherjee B, Park SK. Associations between repeated measures of urinary phthalate metabolites with hormones and timing of natural menopause. J Endocr Soc. 2023;7:bvad024.
57. Centers for Disease Control and Prevention (CDC). National diabetes statistics report; 2022. https://www.cdc.gov/diabetes/data/statistics-report/index.html. Accessed 30 Jun 2025.
58. WHO. Obesity and overweigh; 2024. https://www.who.int/news-room/fact-sheets/detail/obesity-and-overweight. Accessed 30 Jun 2025.
59. Emmerich S, Fryar C, Stierman B, Ogden C, (U.S.) NCfHS. Obesity and severe obesity prevalence in adults. United States, August 2021–August 2023. Hyattsville, MD; 2024.
60. Hirode G, Wong RJ. Trends in the prevalence of metabolic syndrome in the United States, 2011–2016. JAMA. 2020;323:2526–8.
61. Cornier MA, Dabelea D, Hernandez TL, Lindstrom RC, Steig AJ, Stob NR, et al. The metabolic syndrome. Endocr Rev. 2008;29:777–822.
62. Damcott CM, Sack P, Shuldiner AR. The genetics of obesity. Endocrinol Metab Clin North Am. 2003;32:761–86.
63. Vaag A, Poulsen P. Twins in metabolic and diabetes research: what do they tell us? Curr Opin Clin Nutr Metab Care. 2007;10:591–6.
64. Chen JQ, Brown TR, Russo J. Regulation of energy metabolism pathways by estrogens and estrogenic chemicals and potential implications in obesity associated with increased exposure to endocrine disruptors. Biochim Biophys Acta. 2009;1793:1128–43.
65. Heindel JJ, Blumberg B, Cave M, Machtinger R, Mantovani A, Mendez MA, et al. Metabolism disrupting chemicals and metabolic disorders. Reprod Toxicol. 2017;68:3–33.
66. Thayer KA, Heindel JJ, Bucher JR, Gallo MA. Role of environmental chemicals in diabetes and obesity: a national toxicology program workshop review. Environ Health Perspect. 2012;120:779–89.
67. Nadal A, Quesada I, Tudurí E, Nogueiras R, Alonso-Magdalena P. Endocrine-disrupting chemicals and the regulation of energy balance. Nat Rev Endocrinol. 2017;13:536–46.
68. Park SK, Wang X, Ding N, Karvonen-Gutierrez CA, Calafat AM, Herman WH, et al. Per- and polyfluoroalkyl substances and incident diabetes in midlife women: the Study of Women's Health Across the Nation (SWAN). Diabetologia. 2022;65:1157–68.
69. Ding N, Karvonen-Gutierrez CA, Mukherjee B, Calafat AM, Harlow SD, Park SK. Per- and polyfluoroalkyl substances and incident hypertension in multi-racial/ethnic women: The Study of Women's Health Across the Nation. Hypertension. 2022;79:1876–86.
70. Ding N, Karvonen-Gutierrez CA, Herman WH, Calafat AM, Mukherjee B, Park SK. Perfluoroalkyl and polyfluoroalkyl substances and body size and composition trajectories in midlife women: the study of women's health across the nation 1999–2018. Int J Obes (Lond). 2021;45:1937–48.
71. Kang H, Ding N, Karvonen-Gutierrez CA, Mukherjee B, Calafat AM, Park SK. Per- and polyfluoroalkyl substances (PFAS) and lipid trajectories in women 45–56 years of age: The Study of Women's Health Across the Nation. Environ Health Perspect. 2023;131:87004.
72. Lee S, Karvonen-Gutierrez C, Mukherjee B, Herman WH, Harlow SD, Park SK. Urinary concentrations of phenols and parabens and incident diabetes in midlife women: The Study of Women's Health Across the Nation. Environ Epidemiol. 2021;5:e171.

73. Nowak K, Ratajczak-Wrona W, Górska M, Jabłońska E. Parabens and their effects on the endocrine system. Mol Cell Endocrinol. 2018;474:238–51.
74. Karvonen-Gutierrez CA, Park SK, Kim C. Diabetes and menopause. Curr Diab Rep. 2016;16:20.
75. Environmental Working Group (EWG). Teen girls' body burden of hormone-altering cosmetics chemicals; 2008. https://www.ewg.org/research/teen-girls-body-burden-hormone-altering-cosmetics-chemicals. Accessed 30 Jun 2025.
76. Bienkowski B. Starbucks will eliminate all PFAS in its packaging: Environmental healthnews; 2022. https://www.ehn.org/starbucks-pfas. Accessed 30 Jun 2025.
77. Reiley L. Major restaurant chains commit to eliminating 'forever chemicals': The Washington Post; 2022. https://www.washingtonpost.com/business/2022/03/24/fast-food-pfas-forever-chemicals/. Accessed 30 Jun 2025.
78. Toxic-Free Future. Retailers committing to phase out PFAS as a class in food packaging and products: Toxic-Free Future; 2024. https://toxicfreefuture.org/mind-the-store/retailers-committing-to-phase-out-pfas-as-a-class-in-food-packaging-and-products/. Accessed 30 Jun 2025.
79. Cordner A, Goldenman G, Birnbaum LS, Brown P, Miller MF, Mueller R, et al. The true cost of PFAS and the benefits of acting now. Environ Sci Technol. 2021;55:9630–3.
80. Knox KE, Schwarzman MR, Rudel RA, Polsky C, Dodson RE. Trends in NHANES biomonitored exposures in California and the United States following enactment of California's Proposition 65. Environ Health Perspect. 2024;132:107007.

Part V
Innovative Approaches and Emerging Technologies

Machine Learning Applications for Brain Disorders: Towards Inclusivity and Equity

KongFatt Wong-Lin, Paula L. McClean, and Daniela Tropea

Abstract Brain disorders are generally complex and are associated with many challenges, including inclusivity and equity issues. Here, we discuss some of the challenges, and show, with dementia as case study, that machine learning, especially in its more practical manifestations, can offer more standardised, consistent, objective, inclusive, equitable, and efficient assessments and decision-making.

Keywords Dementia · Alzheimer's disease · Artificial intelligence (AI) · Practical machine learning · Decision support system

1 Introduction

Brain disorders come in different forms, and they can inflict enormous social and economic burden [1–4]. The medical field categorises brain disorders into neuropsychiatric and neurological types [5]. Within neuropsychiatric disorders, there are several types, such as anxiety disorder, major depressive disorder, schizophrenia, psychosis, bipolar disorder, addiction and seizures. Examples of neurological disorders are dementia, Parkinson's disease, amyotrophic lateral sclerosis (ALS) (motor neuron disease) and multiple sclerosis (MS), which are neurodegenerative, and attention deficit hyperactivity disorder (ADHD), autism spectrum disorder (ASD), Rett

---

K. Wong-Lin (✉)
Intelligent Systems Research Centre, School of Computing, Engineering and Intelligent Systems, Ulster University, Derry-Londonderry, Northern Ireland, UK
e-mail: k.wong-lin@ulster.ac.uk

P. L. McClean
Personalised Medicine Centre, School of Medicine, Ulster University, Derry-Londonderry, Northern Ireland, UK

D. Tropea
Department of Psychiatry, Trinity Translational Medicine Institute, and Trinity College Institute of Neuroscience, Trinity College Dublin, Dublin, Republic of Ireland

FutureNeuro Research SFI Research Centre, Dublin, Republic of Ireland

H. Lee et al. (eds.), *Sex, Gender, and Emerging Technology in Healthcare: Mitigating Bias and Fostering Equity*, https://doi.org/10.1007/978-981-95-2070-1_9

syndrome, cerebral palsy, intellectual disability and specific learning disability and epilepsy, which are neurodevelopmental disorders.

The distinction among brain disorders may not necessarily be clear cut. For instance, epilepsy can be due to developmental causes and associated with psychiatric (e.g. hallucinatory) symptoms [6]. Having multiple severe depressive episodes could increase the likelihood of dementia later in life [7], and epilepsy, dementia and depression and their treatments can be related [8, 9]. Moreover, disorders such as dementia, associated with a decline in memory and other cognitive abilities, can be caused by a variety of neurodegenerative diseases, with Alzheimer's disease constituting the largest proportion [10]; there can also be mixture of multiple dementia types [11].

The complexity of brain disorders and their associated comorbidities presents immense challenges for brain disorder research and care, particularly in clinical diagnosis, risk factors and prevention [1, 4, 12]. As we shall discuss later, artificial intelligence, and more specifically, machine learning, offers significant potential as a tool to handle the complexity of brain disorders.

Adding to the complexity is the emergent understanding on the impact of gender and sex on brain sciences and disorders. Not only are there known gender- and sex-based differences in brain structure and function [13], but certain brain disorders are more strongly associated with men or women [14–16]. For instance, Rett's syndrome primarily affects females, and depression and Alzheimer's disease are higher in women than men [16]. In contrast, there are substantially more males than females diagnosed with specific learning disability, ADHD and ASD [17–19] In fact, recent studies show people with ASD are more likely to be gender diverse, i.e. gender dysphoria [20].

Many of such gender/sex-specific findings were only recently uncovered or studied. There are several reasons for this. One of them is that the sciences, including brain sciences, have historically been dominated by male researchers [21]. Another reason is that historically, brain and behavioural sciences favoured the study of male animals in experiments, with the concern that hormone cycle in females can cause behavioural variation and the assumption that results from males apply to females [22]. As preclinical drug studies relied mainly on rodent experiments, using only male animals may lead to drug complications in humans [23]. From a clinical perspective, women were historically excluded from participation in phase 1 and phase 2 clinical trials and remain significantly underrepresented in randomised controlled trials [24, 25]. This has led to harm as many drugs have been approved on the basis of trials conducted in men. Adverse drug reactions (ADRs) affect more women than men and sex differences in pharmacokinetics are strongly linked to differences in ADRs [26].

It is only relatively recently that formal policies have required the balancing of the numbers of males and females when designing and executing experiments, and data analysis [27]. The above issues can also have ramifications to machine learning algorithms and models, which are trained on existing data—skewed data will lead to skewed algorithms for diagnostic and prognostic decision-making [28]. Given the significant heterogeneity within, and overlap between, neuropsychiatric and neurological conditions, many factors may contribute to risk and prognosis and methods applied for such analyses must be suitable for the scientific question posed [29].

In addition to the lack of gender inclusivity, there is also health inequity. For instance, differences in ethnicity and cultures may lead to different chances for diagnosis, which could involve higher stigma associated with brain disorders and hence more reluctance to seek medical assistance [30, 31]. Moreover, people from lower income or rural areas will have higher difficulty accessing quality healthcare and treatment [32, 33]. This is exacerbated in countries such as the UK, where there are long delays in clinical consultations and assessments (e.g. see Wong-Lin et al. [34] for dementia care pathway).

It is often supposed that more advanced technologies, that come with higher costs and often based in urban areas, are associated with higher diagnostic accuracy [35]. An example is the use of positron emission tomography (PET) and magnetic resonance imaging (MRI) technologies for brain scans to identify brain disorders. These require relatively costly machines to operate and are often located in hospitals and medical centres in cities. Can machine learning algorithms and models assist in identifying diagnostic assessments that do not require expensive assessments, yet not sacrificing diagnostic accuracy, thus improving equity? This and other challenges will be addressed in the rest of the chapter, with focus on dementia due to its complex etiologies, prolonged degeneration, lack of reliable and robust (bio)markers for detection and prediction, poor accuracy (misdiagnosis) and inconsistency in diagnosis/prognosis, gender influences, pervasiveness and growing trend, and inequity [36, 37].

2 Machine Learning Applications for Brain Disorders

Some of the previously mentioned challenges around equity and accessibility could be ameliorated with machine learning. Machine learning, a sub-field of artificial intelligence, generally consists of the process of data preparation and preprocessing, model training, model validation and model testing (Fig. 1). Data preparation and preprocessing could involve sorting, cleaning, labelling, filtering, augmenting and normalisation of data before analysis is performed [38–40]. A model or set of models is then appropriately selected and trained on a subset of the data [39]. The choice of the training data is particularly important as it can create significant bias in the discovery [28]. It is important to consider whether the variables are independent, and to make sure that the data have been acquired with comparable tools. It is also important to check that the training data are inclusive of a variety of population.

In supervised learning, the model parameters are iteratively modified such that the model outputs gradually match closer towards that of the desired outputs or targets via validation [39]. In other words, the algorithms seek to minimise the differences between the outputs and targets, i.e. minimise some "loss function". For example, the model could be a simple decision tree [39] and the target could be clinicians' diagnosis. During the testing stage, the trained model is then evaluated on

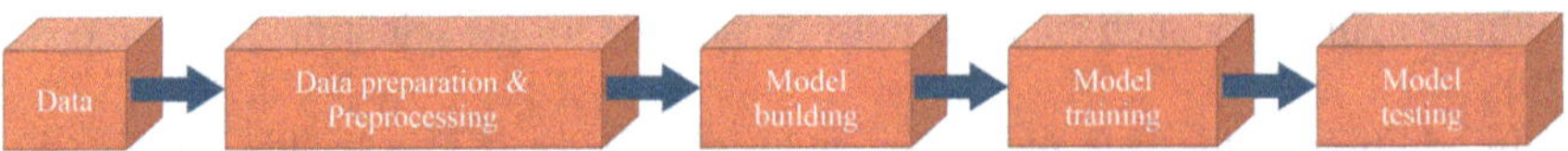

Fig. 1 Standard data handling and modelling process in machine learning. Process begins from data collation (left) through computational model building/selection and training (middle), and testing (right). There is often evaluation of multiple computational models' performance or interpretation prior to deployment

an "unseen" subset of the collated data (removed from the training data earlier during the preparation stage). This process could be evaluated repeatedly and rigorously for example via cross-validation in which the data is divided into multiple "folds" for repeated computations while mitigating sampling bias [39]. Critically, the process is highly automated, with several orders of magnitude faster than manual or statistical computations and can reduce human biases in the decision- making process.

Another often used machine learning approach is unsupervised learning, in which patterns (e.g. clusters or associations) in the data can be identified without *a priori* hypotheses and class labels (or bias) [39]. For example, targets or labels could be associated with the individual data clusters instead of conforming to human labelling, and hence, it is a more data-driven machine learning approach.

Decision-making in machine learning for diagnosis or prognosis is carried out during model building, training and testing, often which classification on discrete classes (e.g. detecting dementia from healthy controls) or regression over continuous outcomes (e.g. dementia severity level) is involved [34, 39, 41]. Hence, the diagnostic process can in some ways be automated. By automatically identifying a smaller subset of key factors through the methods of feature selection or feature extraction [39], machine learning can handle and provide deeper insights into key data features associated with specific brain disorders [41]. Thus, machine learning could potentially address the lack of standardisation, consistency, objectivity and efficiency in clinical assessments [34].

Despite the promising applications of machine learning, several challenges have been identified that influence the translational potential of machine learning applications for brain disorders. Ahmed et al. [42] discusses significant ethical, technological, liability and regulatory, workforce, social, and patient safety barriers that challenge the adoption of novel algorithms within routine healthcare settings, which we expand upon here.

First, most machine learning models are not readily adopted by clinicians or clinical practice, remaining within the confines of "academic" communities. Making the algorithms more interpretable and user friendly to promote adoption by clinicians could be potential solutions [43]. In terms of inclusivity, clinical decision support systems built on machine learning models should involve and include more "humans-in-the-loop" processes for co-designing [44], e.g. involving clinicians to fine tune the models during training. Researchers in the field must take note of relevant guidelines, such as those released by the UK Government [45] which illustrates the requirements for commissioning such a took in healthcare practice and associated Evidence Standard Frameworks [46]. However, this is not often the approach followed in most research studies.

Second, most clinical data are heterogeneous, that is of mixed data types. For instance, they may include brain imaging, blood-related data, medical or family history, medication history and cognitive assessments. The earliest and most common area for machine learning application in dementia is in the subfield of computational neuroimaging, due to the well-known discriminative power of modern AI in imaging data, especially deep learning algorithms (extensive neural network models mimicking brain computation) [34]. For example, deep learning on MRI data could distinguish Alzheimer's disease patients from normal healthy people with very high accuracies (outperforming clinicians), and the model trained in one dataset can be used for other datasets. Functional brain changes have also been investigated using machine learning [34, 47]. However, non-imaging data are not often subject to deep learning studies.

Further, actual clinical data could be "dirty", with a significant proportion of missing data or poor labelling [48], and there needs to be extensive use of data cleaning and preprocessing approaches when dealing with such data. In fact, we have indicated that poor data preprocessing and preparation may lead to inconsistency in dementia prognoses or risk analyses [49]. In particular, utilising available longitudinal diagnostic data, as opposed to a baseline snapshot, as is the case for a majority of published risk factor studies, increases reliability and reproducibility of findings.

Data imputation of missing data can also be performed using machine learning, even with large proportion of missingness [50–52]. Data imputation can be performed while evaluating the overall computational time of the imputation; we showed that imputation algorithms vary more widely in terms of computational time than Alzheimer's disease classification accuracy [51]. Missing data may be more prevalent and more important in more rural areas due to difficulty accessing healthcare [53]. It is also more prevalent in diseases such as dementia with long period of degeneration and long periods of follow ups, leading to higher dropouts of study cohorts [54, 55].

Third, assessments of brain disorders may not be optimised for efficiency. For example, questionnaires or assessments of various cognitive and functional assessments may overlap [56]. Some of these assessments such as the mini-mental state examination (MMSE) are decades old without much updates [57]. Further, it is unclear what combination of assessments gives the most accurate diagnosis or prognosis for specific brain disorders. Machine learning may be useful in this case using feature selection or extraction techniques. For example, iterative algorithms on various assessments could be rapidly combined and their predictions computed, and the most accurate combinations could be automatically determined [58, 59]. However, sub- assessments (e.g. specific questions) are often not incorporated and evaluated. In terms of efficiency, the time duration of (sub)assessments and financial costs have often not been evaluated as part of the feature selection or extraction procedures. Hence, rendering them less practical or less inclusive (for financially poorer patients or under resourced hospitals and medical centres).

Fourth, supervised learning machine-learning algorithms applied to healthcare and biomedical data often assume clinical diagnosis to be the "gold standard". These are also called classes, labels or targets in supervised learning wherein the outputs of the machine learning models aim to reach towards these targets by changing the models' parameters iteratively. However, clinician diagnosis may not always be correct. For example, it is known that a substantial proportion of UK general practitioners are not confident with diagnosing dementia [60]. Indeed, different barriers exist across continents and regions that influence diagnosis, spanning sociodemographic, cultural and physician training that must be addressed [61]. Hence, utilisation of clinician diagnostic labels to train models may be imprecise.

Fifth, there is still insufficient investigation on the roles of gender/sex and ethnicity on specific brain disorders. For example, many of the brain imaging templates still use a Caucasian head type [62]. More generally, research studies on brain sciences and disorders are generally focused on Caucasians [63], even though there are known differences in disease susceptibility attributable to single nucleotide polymorphism (SNPs) within ethnic groups [64]. Beyond biological related studies, there are also social and cultural aspects such as self and societal perception of brain disorders can vary widely [65]. Therefore, more diverse data is needed to better understand brain disorders.

3 Towards Inclusive and Equitable Machine Learning for Brain Disorders

To demonstrate how machine learning can be made more inclusive and equitable, we shall provide case studies on dementia, and particularly Alzheimer's disease. First, to encompass a more holistic machine learning approach, machine learning needs to embrace heterogeneity. For example, to handle highly heterogeneous data and to evaluate dynamical changes in relationship of data features, probabilistic causal Bayesian network modelling on longitudinal data of time-evolved features and assessments can be conducted at baseline and follow up several years later [58].

In Fig. 2a, b, the Bayesian network model identified the most important relationship with respect to detecting healthy controls, people with Alzheimer's disease or those in its prodromal stage, mild cognitive impairment. The data used included demographics (age), brain imaging (MRI and PET), and cognitive and functional assessments (clinical dementia rating, CDR; MMSE; logical memory immediate recall, LMIR; and logical memory delayed recall, LMDR). From the graphical network, one could visualise how the relationship between these time-changing variables dynamically changed over the years.

Using CDR as an objective measure of dementia stage, the model constructed showed the strongest relationship with MMSE, followed by LMDR, then LMIR, indicating that cognitive and functional assessments were the most important features determining dementia stage identification. Using predisposing indicators/biomarkers

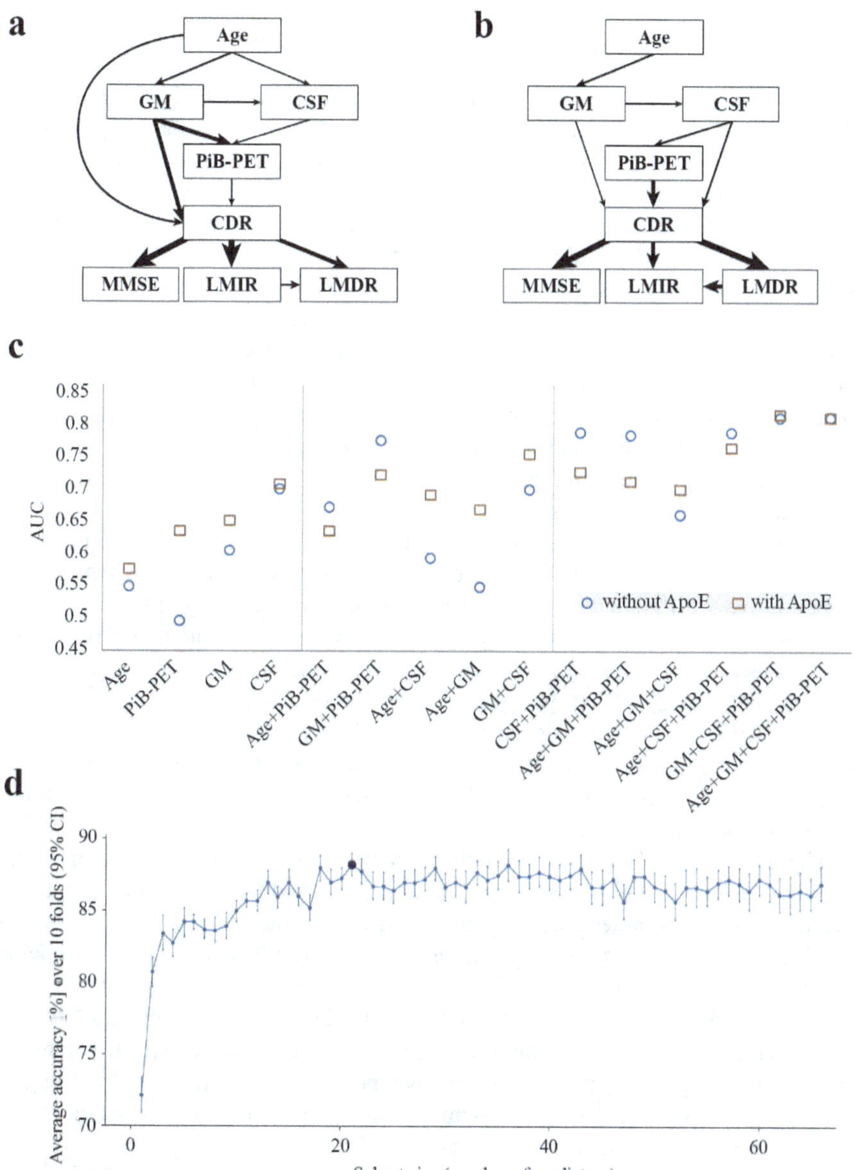

Fig. 2 Feature selection and extraction on heterogeneous Alzheimer's disease data. **a**, **b** Bayesian network at baseline (**a**) and years later (**b**), indicating the relationship between variables had dynamically changed. Arrow thickness: strength of probabilistic causal influences between variables. **c** Accuracy based on area under the receiver operating characteristics (ROC) curve (AUC) (adapted from Ding et al. [58]). **d** Recursive feature elimination random forest technique used for identifying the optimal subset of combined features (i.e. assessments). Filled circle: optimal subset size (adapted from Bucholc et al. [59])

and their direct/indirect influences on the CDR score achieved high multi-class classification accuracy and AUC score.

Figure 2c shows the model constructed using individual as well as combinations of predisposing factors/biomarkers with and without genetic risk factor apolipoprotein E (*ApoE*) alleles. Although the incorporation of *ApoE* alleles generally improved model performance, other combinations did not necessarily lead to large drop in classification accuracy. Thus, generally, there are different combinations of assessments that could attain similarly high accuracy, and cognitive and functional assessments were found to be key assessment metrics.

Another of our studies [59] on a different heterogeneous data made use of feature selection to understand the influence of different feature subset sizes on identifying an individual's Alzheimer's disease stage (Fig. 2d). We obtained the optimal subset of features (assessments) that gave the highest classification accuracy. Again, in this study, cognitive and functional assessments were found to be the strongest predictors of disease stage, outperforming or on par with coarse-grained (volumetric-based) level neuroimaging data types.

In another study, we used simple interpretable machine learning (associative rule mining) algorithms to distinguish Alzheimer's disease from Lewy body dementia (two different types of dementia), attaining very high accuracy, sensitivity and specificity, while identifying some key variables, including one (limb rigidity) not used in clinical practice [66]. We built models that included ^{123}I-2β-carbomethoxy-3β-(4-iodophenyl)-N-(3-fluoropropyl) nortropane SPECT (FP-CIT) and ^{123}I-metaiodobenzylguanidine (MIBG) cardiac scintigraphy imaging and without imaging options, with the latter having comparable accuracies. Thus, basic clinical measurements may be utilised to avoid the high costs and long waiting times to use specialist imaging facilities.

These and related studies indicate that machine learning can assist in identifying smaller number of data variables and assessments without sacrificing diagnostic accuracy. Some variables (e.g. cognitive assessment) have relatively lower costs to administer, and could be simply implemented, e.g. via a mobile app, for wide accessibility—reducing inequity.

Such machine learning approach can be extended to incorporate sub-assessments, e.g. selecting only the key questions in a questionnaire (Fig. 3a, b) [67]. The resultant selected battery of (sub)assessments may potentially lead to the redesigning of more efficient diagnostic assessments and shorten clinical appointments, potentially reducing waiting times in an overburdened service.

For more practical applications, feature selection can even incorporate time and financial costs, as part of a "penalty" term within standard loss function for algorithmic optimisation, a kind of regularisation [67, 68]. Figure 3c shows the combinations of all possible subsets of assessments based on diagnostic accuracy, time cost and total financial cost, offering flexibility of choice based on specific needs. For instance, if there is personal financial limitation, one can opt for a subset of assessments with lower financial cost without sacrificing diagnostic accuracy (Fig. 3c, orange data point towards the top). This can lead to improved equity in dementia care.

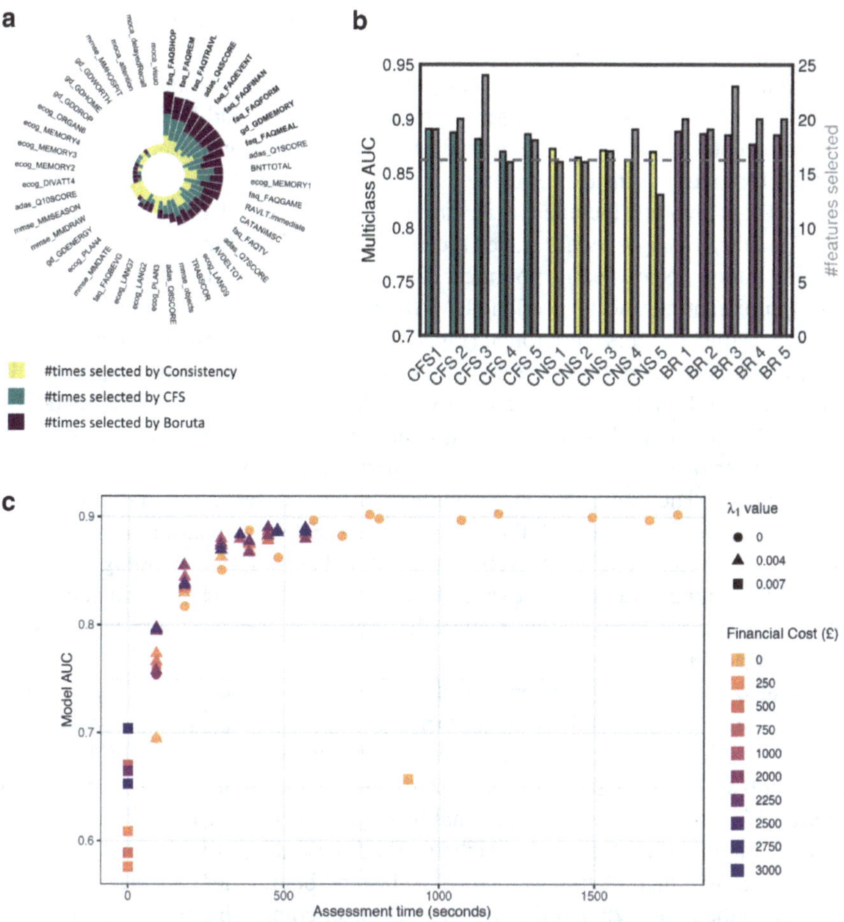

Fig. 3 Feature selection based on dementia sub-assessments (specific questions in questionnaire) and costs. **a** Three feature selection methods (three colours) used on individual (sub)assessments (names on circles). **b** High Alzheimer's disease classification accuracy for subsets of features, even outperforming that using original full dataset (dashed line). Right vertical label: number of features selected by a feature selection method. Colour labelling as in (**b**) (adapted from McCombe et al. [67]). **c** Evaluating combination of assessment features while considering their total time duration of assessment and total financial costs (adapted from McCombe et al. [68])

Clinicians and other stakeholders can also be involved in the process of data building or development—a "humans-in-the-loop" approach. Built on top of our previous work, we have developed machine learning applications with a user-friendly interface (via the Shiny package in R) for identifying not only key features for identifying dementia, but also allows non-technical users such as clinicians to directly select the features (based on their local medical practice) and train the model, leading towards enhancing user engagement while incorporating their domain expertise and knowledge [67, 68]. This enhances inclusivity in modelling and subsequently may improve usability and accuracy.

Given that clinician diagnosis of brain disorders such as Alzheimer's disease diagnosis may not be consistent or highly accurate (e.g. Koch and Iliffe [60]), we have trained machine learning models with data-driven unsupervised learning to identify clusters for labelling instead of using clinician's diagnosis [69]. The highly heterogeneous data involved included active brain regions from PET imaging for detecting tau pathology (using specific tracer recently approved by the US Food and Drug Administration). Specifically, nonlinear dimensional reduction using uniform manifold approximation and projection (UMAP) and k-means clustering were applied and five distinct clusters were found (Fig. 4a, b). One cluster comprised mainly clinically diagnosed Alzheimer's disease cases, two other clusters each comprised fully either males or females, a fourth cluster which had almost equally mixed gender but with a substantially high number of control normal cases with neurological risks, and a fifth cluster of younger participants with parental history of Alzheimer's disease (a risk factor). The minority non-Alzheimer's disease cases in the first cluster have intermediate values for tau-PET in the amygdala and entorhinal brain regions and memory assessment, and their cerebral spinal fluid beta amyloid pathology (not used in model training) was more advanced (lower value) than in the clinically diagnosed Alzheimer's disease cases. Hence, the clusters offered disease risk predictions without *a priori* hypotheses.

Using feature selection, the top ranked features highly associated with Alzheimer's disease stage were found to be APoE4, gender, and history of parents with Alzheimer's disease, followed by certain active brain regions from tau-PET imaging. The results were consistent with a later study using a different approach on the same dataset [52]. Hence, gender and parental history of the disease (especially mother's history) are important factors for Alzheimer's disease diagnosis. Importantly, clinician's diagnosis did not align well with the clusters' boundaries (Fig. 4c, d). We then re-labelled the non-Alzheimer's cases in the cluster comprising mainly Alzheimer's disease cases as Alzheimer's disease, and then trained a deep learning (graph neural network) model with it. We found the deep learning model with the new labels to have better diagnostic performance than that using clinician's diagnosis as training/testing labels. This illustrates how data-driven machine learning approaches can uncover key data features for disease diagnosis and risk predictions, including consideration of gender-based factors. This may also in turn leads to less human biases (e.g. regardless of clinician experience) and enhances objectivity in brain disorder studies.

Given the significant, and expanding, Alzheimer's disease population worldwide, and the lack of effective, accessible therapeutic options, significant efforts are being made to identify novel, disease modifying drugs. Alzheimer's disease clinical trials are associated with a considerably higher failure rate than other indications (99% vs. 90%) [70, 71], which has been attributed to a range of factors including the complex pathophysiology of disease, the heterogeneity of the patient population, the long prodrome before overt memory deficits emerge, interventions being "too late" and potential misdiagnosis [72]. The anti-amyloid antibody-based drugs, Aducanumab [73] and Lecanemab [74] have only gained regulatory approval in the US and their

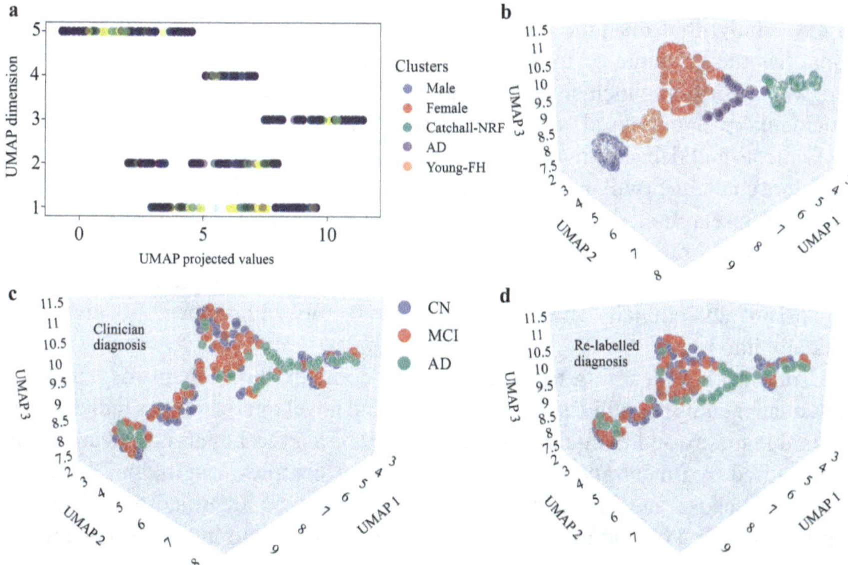

Fig. 4 Data-driven nonlinear dimensional reduction and clustering reveals key Alzheimer's disease factors, and for re-labelling for enhanced classification. **a** Data projection on UMAP space. Colour labelling of classes (clinician diagnosis): control normal, CN (dark purple), mild cognitive impairment, MCI (light purple), and Alzheimer's disease, AD (yellow). **b** Five distinct clusters in lower three-dimensional UMAP space reveals key data features. **c** Clinician diagnosis (CN) not well-aligned with cluster boundaries. **d** Re-labelling of CN and MCI data points in cluster comprising mainly AD cases to AD cases (adapted from McCombe et al. [69])

cost and lack of approval worldwide limits access for patients. Accordingly, increased emphasis has been placed on utilising computational drug repurposing strategies in the context of Alzheimer's disease to address barriers and limitations associated with traditional drug development strategies. We propose that combining computational drug repurposing strategies and a stratified medicine approach, initially focused on sex, genotype and age, will identify better, more targeted therapies that will address cost, availability, and time-to-market-related barriers associated with traditional strategies [72].

4 Outlook and Further Challenges

We have discussed the various challenges facing brain disorders, including those affecting inclusivity and equity. We have also provided brief introduction on how machine learning can process complex, large, high-dimensional, heterogeneous and "dirty" clinical data associated with brain disorders. We have shown, with dementia as case study, that machine learning, especially in its more practical manifestations, has the potential to transform the diagnosis and prognosis of brain disorders while enhancing inclusivity and equity. Thus, machine learning can potentially provide more standardised, consistent, objective, inclusive, equitable, and efficient assessments and decision-making.

Future challenges will involve including more gender and ethnically diverse populations in research studies, clinical validation and adoption of algorithms, and in distinguishing brain disorder subtypes and co-morbidities. Personalised medicine approaches to evaluate subtypes within each disorder and simple stratification on the basis of sex, ethnicity and core genetic risk factors within novel models will be critically important.

Using machine learning to understand the complex biomechanisms leading to particular phenotypes will also uncover causes and novel treatments, including differences due to sex and ethnicity. These will in turn lead to better stratification, and personalised treatment and care. Through mining complex longitudinal data, risk factors and causal mechanisms of brain disorders can be identified using machine learning, which will lead to better preventative measures and interventions. Finally, higher quality data, including having more inclusive and diverse data, requires improved funding, data acquisition, integration and curation, and with appropriate practical machine learning methodologies, may lead to data-driven re-designing of care pathways for brain disorders.

Acknowledgements K. W.-L. wish to thank H. Lee (editor; GISTeR President) for the invitation to present at the 2023 BrainLink X-Lab Day Exploring Life Phenomena: Integrating Clinical Science and Artificial Intelligence for Inclusive Research, Pyeongchang and Seoul, Republic of Korea. The content of this chapter was based largely on that presentation. Some of the mentioned studies were previously funded by the European Union's INTERREG VA Programme, managed by the Special EU Programmes Body (SEUPB; Centre for Personalised Medicine, IVA 5036) (K. W.-L., P. L. M.), with additional support by the Northern Ireland Functional Brain Mapping Project Facility (1303/101154803) and Invest Northern Ireland (K. W.-L.), Alzheimer's Research UK (ARUK) NI Pump Priming (K. W.-L., P. L. M.) and Ulster University Research Challenge Fund (K. W.-L., P. L. M.). The views and opinions expressed in this paper do not necessarily reflect those of the European Commission or the Special EU Programmes Body (SEUPB). D.T. is partially supported by IRSF (3507-207417 grant), Meath Foundation (Research award 2019 to D. T. as co-PI), and Science Foundation Ireland (SFI) under Grant Number 16/RC/3948 and co-funded under the European Regional Development Fund and by FutureNeuro industry partners. D. T. also received the Trinity College Dublin School of Medicine Seed Award.

References

1. Feigin VL, Vos T, Nichols E, Owolabi MO, Carroll WM, Dichgans M, et al. The global burden of neurological disorders: translating evidence into policy. Lancet Neurol. 2020;19:255–65.
2. Huang Y, Li Y, Pan H, Han L. Global, regional, and national burden of neurological disorders in 204 countries and territories worldwide. J Glob Health. 2023;13:04160.
3. Winkler AS, Gupta S, Patel V, Bhebhe A, Fleury A, Aukrust CG, et al. Global brain health-the time to act is now. Lancet Glob Health. 2024;12:e735–6.

4. Grisold W. The expanding burden of neurological disorders. Lancet Neurol. 2024;23:326–7.
5. Association AP. Diagnostic and statistical manual of mental disorders, 5th ed. Arlington, VA: American Psychiatric Association; 2013.
6. Vinti V, Dell'Isola GB, Tascini G, Mencaroni E, Cara GD, Striano P, et al. Temporal lobe epilepsy and psychiatric comorbidity. Front Neurol. 2021;12:775781.
7. Byers AL, Yaffe K. Depression and risk of developing dementia. Nat Rev Neurol. 2011;7:323–31.
8. Keezer MR, Sisodiya SM, Sander JW. Comorbidities of epilepsy: current concepts and future perspectives. Lancet Neurol. 2016;15:106–15.
9. Joshi A, Todd S, Finn DP, McClean PL, Wong-Lin K. Multi-dimensional relationships among dementia, depression and prescribed drugs in England and Wales hospitals. BMC Med Inform Decis Mak. 2022;22:262.
10. Gale SA, Acar D, Daffner KR. Dementia. Am J Med. 2018;131:1161–9.
11. Jellinger KA. The enigma of mixed dementia. Alzheimers Dement. 2007;3:40–53.
12. Tropea D, Harkin A. Editorial: biology of brain disorders. Front Cell Neurosci. 2017;11:366.
13. Zhang X, Liang M, Qin W, Wan B, Yu C, Ming D. Gender differences are encoded differently in the structure and function of the human brain revealed by multimodal MRI. Front Hum Neurosci. 2020;14:244.
14. Thibaut F. The role of sex and gender in neuropsychiatric disorders. Dialogues Clin Neurosci. 2016;18:351–2.
15. Bölte S, Neufeld J, Marschik PB, Williams ZJ, Gallagher L, Lai MC. Sex and gender in neurodevelopmental conditions. Nat Rev Neurol. 2023;19:136–59.
16. DuMont M, Agostinis A, Singh K, Swan E, Buttle Y, Tropea D. Sex representation in neurodegenerative and psychiatric disorders' preclinical and clinical studies. Neurobiol Dis. 2023;184:106214.
17. Moll K, Kunze S, Neuhoff N, Bruder J, Schulte-Körne G. Specific learning disorder: prevalence and gender differences. PLoS ONE. 2014;9:e103537.
18. Loomes R, Hull L, Mandy WPL. What is the male-to-female ratio in autism spectrum disorder? A systematic review and meta-analysis. J Am Acad Child Adolesc Psychiatry. 2017;56:466–74.
19. Mowlem FD, Rosenqvist MA, Martin J, Lichtenstein P, Asherson P, Larsson H. Sex differences in predicting ADHD clinical diagnosis and pharmacological treatment. Eur Child Adolesc Psychiatry. 2019;28:481–9.
20. Thrower E, Bretherton I, Pang KC, Zajac JD, Cheung AS. Prevalence of autism spectrum disorder and attention-deficit hyperactivity disorder amongst individuals with gender dysphoria: a systematic review. J Autism Dev Disord. 2020;50:695–706.
21. Halpern DF, Benbow CP, Geary DC, Gur RC, Hyde JS, Gernsbacher MA. The science of sex differences in science and mathematics. Psychol Sci Public Interest. 2007;8:1–51.
22. Beery AK, Zucker I. Sex bias in neuroscience and biomedical research. Neurosci Biobehav Rev. 2011;35:565–72.
23. Rademaker M. Do women have more adverse drug reactions? Am J Clin Dermatol. 2001;2:349–51.
24. Seydel C. The missing sex. Nat Biotechnol. 2021;39:260–5.
25. National Academies of Sciences E, and Medicine, Affairs PaG, Committee on Women in Science E, and Medicine, Research ColtRoWaUMiCTa, Bibbins-Domingo K, Helman A. Improving representation in clinical trials and research: building research equity for women and underrepresented groups. Washington, DC: The National Academies Press; 2022. 280 p.
26. Zucker I, Prendergast BJ. Sex differences in pharmacokinetics predict adverse drug reactions in women. Biol Sex Differ. 2020;11:32.
27. Wierenga LM, Ruigrok A, Aksnes ER, Barth C, Beck D, Burke S, et al. Recommendations for a better understanding of sex and gender in the neuroscience of mental health. Biol Psychiatry Glob Open Sci. 2024;4:100283.
28. Obermeyer Z, Emanuel EJ. Predicting the future—big data, machine learning, and clinical medicine. N Engl J Med. 2016;375:1216–9.

29. Feczko E, Fair DA. Methods and challenges for assessing heterogeneity. Biol Psychiatry. 2020;88:9–17.
30. Anglin DM, Link BG, Phelan JC. Racial differences in stigmatizing attitudes toward people with mental illness. Psychiatr Serv. 2006;57:857–62.
31. Abdullah T, Brown TL. Mental illness stigma and ethnocultural beliefs, values, and norms: an integrative review. Clin Psychol Rev. 2011;31:934–48.
32. Mseke EP, Jessup B, Barnett T. Impact of distance and/or travel time on healthcare service access in rural and remote areas: a scoping review. J Transp Health. 2024;37: 101819.
33. Dawkins B, Renwick C, Ensor T, Shinkins B, Jayne D, Meads D. What factors affect patients' ability to access healthcare? An overview of systematic reviews. Trop Med Int Health. 2021;26:1177–88.
34. Wong-Lin K, McClean PL, McCombe N, Kaur D, Sanchez-Bornot JM, Gillespie P, et al. Shaping a data-driven era in dementia care pathway through computational neurology approaches. BMC Med. 2020;18:398.
35. Rong J, Liu Y. Advances in medical imaging techniques. BMC Methods. 2024;1:10.
36. Dubois B, Hampel H, Feldman HH, Scheltens P, Aisen P, Andrieu S, et al. Preclinical Alzheimer's disease: definition, natural history, and diagnostic criteria. Alzheimers Dement. 2016;12:292–323.
37. Prince MJ, Comas-Herrera A, Knapp M, Guerchet MM, Karagiannidou M. World Alzheimer Report 2016—improving healthcare for people living with dementia: coverage, quality and costs now and in the future. Alzheimer's Disease International; 2016.
38. Pyle D. Data preparation for data mining. Los Altos, California: Morgan Kaufmann Publishers; 1999.
39. Hastie T, Tibshirani R, Friedman JH. The elements of statistical learning: data mining, inference, and prediction. New York: Springer; 2009.
40. García S, Luengo J, Herrera F. Data preprocessing in data mining, vol. 72. Cham, Switzerland: Springer International Publishin; 2015.
41. Wong-Lin K, Sanchez-Bornot J, McCombe N, Kaur D, McClean PL, Zou X, et al. Computational neurology: computational modeling approaches in dementia. In: Wolkenhauer O, editor. Systems medicine: integrative, qualitative and computational approaches, vol. 2. Chantilly: Elsevier Science & Technology; 2020. p. 81–9.
42. Ahmed MI, Spooner B, Isherwood J, Lane M, Orrock E, Dennison A. A systematic review of the barriers to the implementation of artificial intelligence in healthcare. Cureus. 2023;15: e46454.
43. Doshi-Velez F, Kim B. Towards a rigorous science of interpretable machine learning. arXiv: Machine Learning; 2017.
44. Amershi S, Weld DS, Vorvoreanu M, Fourney A, Nushi B, Collisson P, et al. Guidelines for human-AI interaction. In Proceedings of the 2019 CHI conference on human factors in computing systems. Glasgow, Scotland Uk: Association for Computing Machinery; 2019. p. Paper 3.
45. Care DoHaS. A guide to good practice for digital and data-driven health technologies. In: Care DoHaS, editor. GOV.UK; 2021.
46. (NICE) NIfHaCE. Evidence standards framework for digital health technologies; 2019.
47. Yang S, Bornot JMS, Wong-Lin K, Prasad G. M/EEG-based bio-markers to predict the MCI and Alzheimer's disease: a review from the ML perspective. IEEE Trans Biomed Eng. 2019;66:2924–35.
48. Raghupathi W, Raghupathi V. Big data analytics in healthcare: promise and potential. Health Inf Sci Syst. 2014;2:3.
49. Kaur D, Bucholc M, Finn DP, Todd S, Wong-Lin K, McClean PL. Multi-time-point data preparation robustly reveals MCI and dementia risk factors. Alzheimers Dement (Amst). 2020;12: e12116.
50. McCombe N, Ding X, Prasad G, Finn D, Todd S, McClean PL, et al., editors. Predicting feature imputability in the absence of ground truth. In: Proceedings of the 37th international conference on machine learning (ICML): the art of learning with missing values (ARTEMISS) workshop; 2020 17 July; Vienna, Austria.

51. McCombe N, Liu S, Ding X, Prasad G, Bucholc M, Finn DP, et al. Practical strategies for extreme missing data imputation in dementia diagnosis. IEEE J Biomed Health Inform. 2022;26:818–27.
52. Haridas NT, Sanchez-Bornot JM, McClean PL, Wong-Lin K. Autoencoder imputation of missing heterogeneous data for Alzheimer's disease classification. Healthc Technol Lett. 2024;11:452–60.
53. Chen X, Orom H, Hay JL, Waters EA, Schofield E, Li Y, et al. Differences in rural and urban health information access and use. J Rural Health. 2019;35:405–17.
54. Brodaty H, Seeher K, Gibson L. Dementia time to death: a systematic literature review on survival time and years of life lost in people with dementia. Int Psychogeriatr. 2012;24:1034–45.
55. William-Faltaos D, Chen Y, Wang Y, Gobburu J, Zhu H. Quantification of disease progression and dropout for Alzheimer's disease. Int J Clin Pharmacol Ther. 2013;51:120–31.
56. Velayudhan L, Ryu SH, Raczek M, Philpot M, Lindesay J, Critchfield M, et al. Review of brief cognitive tests for patients with suspected dementia. Int Psychogeriatr. 2014;26:1247–62.
57. Folstein MF, Folstein SE, McHugh PR. "Mini-mental state". A practical method for grading the cognitive state of patients for the clinician. J Psychiatr Res. 1975;12:189–98.
58. Ding X, Bucholc M, Wang H, Glass DH, Wang H, Clarke DH, et al. A hybrid computational approach for efficient Alzheimer's disease classification based on heterogeneous data. Sci Rep. 2018;8:9774.
59. Bucholc M, Ding X, Wang H, Glass DH, Wang H, Prasad G, et al. A practical computerized decision support system for predicting the severity of Alzheimer's disease of an individual. Expert Syst Appl. 2019;130:157–71.
60. Koch T, Iliffe S. Rapid appraisal of barriers to the diagnosis and management of patients with dementia in primary care: a systematic review. BMC Fam Pract. 2010;11:52.
61. Kerwin D, Abdelnour C, Caramelli P, Ogunniyi A, Shi J, Zetterberg H, et al. Alzheimer's disease diagnosis and management: perspectives from around the world. Alzheimers Dement (Amst). 2022;14: e12334.
62. Rao NP, Jeelani H, Achalia R, Achalia G, Jacob A, Bharath RD, et al. Population differences in brain morphology: need for population specific brain template. Psychiatry Res Neuroimaging. 2017;265:1–8.
63. Henrich J, Heine SJ, Norenzayan A. The weirdest people in the world? Behav Brain Sci. 2010;33:61–83; discussion-135.
64. Huang T, Shu Y, Cai YD. Genetic differences among ethnic groups. BMC Genomics. 2015;16:1093.
65. Kleinman A. Patients and healers in the context of culture: an exploration of the borderland between anthropology, medicine, and psychiatry, 1 ed. University of California Press; 1980.
66. McCombe N, Joshi A, Finn DP, McClean PL, Roberts G, O'Brien JT, et al. Distinguishing Lewy body dementia from Alzheimer's disease using machine learning on heterogeneous data: a feasibility study. Annu Int Conf IEEE Eng Med Biol Soc. 2022;2022:4929–33.
67. McCombe N, Ding X, Prasad G, Gillespie P, Finn DP, Todd S, et al. Alzheimer's disease assessment osptimized for diagnostic accuracy and administration time. IEEE J Transl Eng Health Med. 2022;10:4900809.
68. McCombe N, Ding X, Prasad G, Finn DP, Todd S, McClean PL, et al. Multiple cost optimisation for Alzheimer's disease diagnosis. Annu Int Conf IEEE Eng Med Biol Soc. 2022;2022:1098–104.
69. McCombe N, Bamrah J, Sanchez-Bornot JM, Finn DP, McClean PL, Wong-Lin K. Alzheimer's disease classification using cluster-based labelling for graph neural network on heterogeneous data. Healthc Technol Lett. 2022;9:102–9.
70. Cummings J, Feldman HH, Scheltens P. The, "rights" of precision drug development for Alzheimer's disease. Alzheimers Res Ther. 2019;11:76.
71. Sun D, Gao W, Hu H, Zhou S. Why 90% of clinical drug development fails and how to improve it? Acta Pharm Sin B. 2022;12:3049–62.
72. Anderson C, Bucholc M, McClean PL, Zhang SD. The potential of a stratified approach to drug repurposing in Alzheimer's disease. Biomolecules. 2023;14:11.

73. Mullard A. Landmark Alzheimer's drug approval confounds research community. Nature. 2021;594:309–10.
74. Harris E. Alzheimer drug lecanemab gains traditional FDA approval. JAMA. 2023;330:495.

Equitable AI in Healthcare: Navigating Sex, Gender, and Intersectional Biases in Diagnostics

Heisook Lee, Sejung Yang, and Yusung Chu

Abstract The integration of Artificial Intelligence (AI) in healthcare offers transformative potential for disease prediction and diagnostics, yet this progress is frequently undermined by the replication and amplification of existing health disparities. This chapter critically examines how AI systems, often trained on historically biased and unrepresentative datasets, perpetuate diagnostic errors—specifically false positives and false negatives—that disproportionately affect women, racial minorities, and individuals with intersecting marginalized identities. We explore foundational concepts such as sex, gender, and intersectionality, alongside the crucial role of social determinants of health, in shaping health outcomes and AI performance. Empirical evidence from cardiovascular disease, kidney disease, and psychiatric diagnostics illustrates the tangible harm caused by the "male-default" in medical understanding and the "black box" nature of many AI models. We propose a multi-layered framework for equitable AI development, encompassing data-centric interventions, fairness-aware algorithmic design, robust human oversight, and comprehensive policy and regulatory governance. By advocating an "equity-by-design" approach that prioritizes sex- and gender-sensitive methodologies and intersectional analysis, this chapter argues that AI can be transformed from a bias amplifier into a powerful instrument for achieving genuinely equitable, transparent, and patient-centered healthcare.

Keywords AI in healthcare · Diagnostic disparities · Gender bias · Racial bias · Intersectionality · Sex-disaggregated data · False positives · False negatives · Algorithmic fairness · Explainable AI (XAI) · Bias mitigation strategies · Health equity

H. Lee (✉)
Korea Center for Gendered Innovations in Science and Technology Research, Seoul, Republic of Korea
e-mail: hslee@gister.re.kr

S. Yang · Y. Chu
Department of Precision Medicine, Yonsei University Wonju College of Medicine, Wonju, Republic of Korea

S. Yang
Department of Medical Informatics and Biostatistics, Graduate School, Yonsei University, Wonju, Republic of Korea

H. Lee et al. (eds.), *Sex, Gender, and Emerging Technology in Healthcare: Mitigating Bias and Fostering Equity*, https://doi.org/10.1007/978-981-95-2070-1_10

1 Introduction: AI's Transformative Potential and Ethical Imperative

Artificial Intelligence (AI) is fundamentally transforming public health and medicine, revolutionizing healthcare delivery, disease prediction, population health management, and patient care [1]. AI systems enhance diagnostic procedures through advanced analysis of medical imaging and sophisticated pattern recognition, often rivaling or even surpassing the accuracy of experienced human clinicians. This has led to numerous regulatory approvals for AI-powered diagnostic devices, underscoring their growing impact [2]. Beyond diagnostics, AI-driven predictive analytics play a pivotal role in disease prevention and early detection. By analyzing vast datasets, AI algorithms can identify subtle patterns and risk factors for various conditions, including diabetes, cardiovascular diseases (CVDs), and different cancers [1]. This capability for early detection improves patient outcomes and contributes to healthcare cost reduction. Furthermore, AI systems optimize triage processes, accelerate definitive diagnoses, offer rapid 'second opinions', and continuously monitor disease progression.

A groundbreaking advancement is the integration of multimodal AI, which combines diverse data types such as text from clinical notes, images like X-rays and MRIs, audio from patient speech or heart/lung sounds, video from gait analysis or surgical footage, and sensor inputs from wearable devices [2]. This technology offers a more comprehensive understanding of complex clinical situations, potentially enabling more accurate diagnoses.

Despite AI's immense potential, the same features that make AI effective—its ability to integrate vast and diverse datasets—also create challenges for human understanding and ethical governance [3]. It is crucial to address the ethical implications and ensure that these advancements benefit all sections of society equitably. The misuse or unethical application of AI can lead to increased disparities and further exacerbate adverse outcomes for socially and economically disadvantaged populations [1]. Health equity in the context of AI applications refers to the fair and just distribution of health technologies and their benefits [1]. A core challenge is that AI risks replicating and accelerating existing gender and other biases that have historically excluded populations from equitable care.

The effectiveness of AI is intrinsically linked to the quality and representativeness of the data it learns from. If the data are biased or unrepresentative, the resulting AI models risk perpetuating and even amplifying existing health disparities [3]. Historical medical datasets often underrepresent women and minority groups due to longstanding inequities in biomedical research, healthcare access, research participation, and data collection practices [3]. This underrepresentation can result in AI models that perform well on average but fail to deliver accurate or safe outcomes for these groups, thereby exacerbating existing disparities rather than alleviating them [3].

Furthermore, traditional AI 'success' metrics, such as overall accuracy, often conceal significant disparities in performance across demographic groups. For example, AI models trained on datasets that are not representative of diverse populations may perform well on average but fail to deliver accurate results for underrepresented groups, such as women or racial minorities. Studies have shown that these models may use demographic shortcuts, resulting in "fairness gaps"—discrepancies in diagnostic accuracy between groups—which can directly harm vulnerable populations [4]. This recognition has led to calls for a redefinition of what constitutes "successful AI" in healthcare, with evaluation frameworks that explicitly prioritize equitable impact alongside technical performance.

The transformative potential of AI in healthcare is inherently conditional on the ethical integrity and representativeness of its foundational data. The very feature that makes AI powerful—its ability to integrate vast and diverse datasets—is also its primary vulnerability if the data is flawed. When biased or unrepresentative data are used, the resulting AI models do not merely reflect existing societal biases; they actively scale and intensify their impact on health outcomes. This means that without proactive and rigorous data governance, innovation in AI could inadvertently become a tool for systemic injustice. Consequently, the focus must shift from simply developing powerful AI to developing responsible and equitable AI as a prerequisite for its beneficial application in healthcare. The pursuit of technical optimization without an explicit and measurable focus on equity metrics is thus fundamentally flawed for healthcare AI. A model can be statistically "accurate" in aggregate but ethically harmful in its differential impact on specific populations. This calls for a fundamental paradigm shift in how AI systems are designed, developed, and evaluated, moving beyond purely computational efficiency to integrate social and ethical impact as core performance indicators. Regulatory bodies, funding agencies, and academic institutions should therefore consider mandating equity-focused evaluation frameworks as standard practice for all healthcare AI.

2 Foundational Concepts for Equitable AI in Health

2.1 Differentiating Sex, Gender, and Intersectionality in Health Outcomes

A nuanced understanding of sex, gender, and intersectionality is paramount for developing equitable AI systems in healthcare, as these factors profoundly influence health outcomes and access to care.

2.1.1 Defining Sex and Gender [5]

Sex refers to a multidimensional biological construct based on a cluster of anatomical and physiological traits, including chromosomes, gene expression, hormone levels and function, and reproductive/sexual anatomy. While typically assigned at birth as female or male, it is crucial to recognize the biological variation in how these attributes are expressed, encompassing intersex variations that do not fit neatly into binary categories. Gender, in contrast, refers to the socially constructed roles, behaviors, expressions, and identities of individuals. It influences how people perceive themselves and each other, how they act and interact, and the distribution of power and resources in society. Gender identity is a person's deeply felt, internal and individual experience of gender, which may or may not correspond to their sex assigned at birth, and is not confined to a binary nor is it static, existing along a continuum and capable of changing over time. There is considerable diversity in how individuals and groups understand, experience, and express gender through the roles they take on, societal expectations, and their relations with others, as well as the complex ways gender is institutionalized in society. Clear differentiation between sex and gender is essential for accurate research, effective interventions, and tracking progress on gender equity initiatives. One recommended solution is the implementation of a two-step data collection process: first, asking about current gender identity, and second, about sex assigned at birth.

2.1.2 Impact of Sex and Gender Differences on Health [6]

Biological and physiological differences between sexes can lead to distinct patterns in disease susceptibility, progression, and responses to treatment, ultimately affecting health outcomes. For example, recent large-scale analyses of observational health datasets have revealed systematic disparities in how women and men experience disease diagnosis, with women consistently being older at initial diagnosis and experiencing longer delays between symptom onset and formal disease recognition across numerous conditions [7].

Healthcare has been historically shaped by pervasive gender bias. For decades, clinical research predominantly focused on male bodies, often excluding women entirely from studies or failing to analyse results by sex. This historical bias has created a medical landscape where women are more prone to misdiagnosis, endure longer diagnostic delays, and receive less effective treatments for conditions that present differently in women, such as CVD and autoimmune disorders [8]. Specifically, women are less likely to present with classic chest pain—the primary symptom for men during heart attacks—instead reporting higher rates of fatigue, dyspnea, nausea, or other forms of pain [9]. AI systems trained predominantly on male datasets may not appropriately weight these 'atypical' symptoms, leading to missed or delayed diagnoses for women. Conditions such as endometriosis, polycystic ovary syndrome (PCOS) [10], autoimmune diseases, and chronic pain disorders are systematically underdiagnosed in women, not due to their rarity, but rather because diagnostic models are poorly attuned to how these conditions manifest across diverse bodies [6].

Beyond biological differences, gender (socially constructed roles) can influence all aspects of health-related behaviors, including access to diagnosis and treatment, often due to hierarchical and unequal power relations. Women and girls frequently encounter greater barriers to healthcare and information stemming from lower decision-making power, restrictions on mobility, fewer financial resources, and discriminatory attitudes from healthcare providers. The 'male-default' is a systemic bias woven into the fabric of data collection and the very framing of medical knowledge itself [11]. This influences how symptoms are categorized, how diseases are understood, and consequently, how clinical decisions are made, as well as how AI systems are trained and ultimately perform [3]. This perpetuates a cycle where women's health differences are perceived as anomalies rather than as essential considerations for comprehensive care [11]. For AI in disease prediction, the challenge extends beyond simply including more data on women; it requires a fundamental re-evaluation and re-framing of medical knowledge and data collection practices to move beyond this male-centric paradigm [12]. This implies a shift in the epistemology of medical AI, ensuring that data is not only diverse but also interpreted and weighted in a gender-sensitive manner, actively challenging the notion of a 'normative' male body [12]. The pervasive "male-default" in medical understanding is not merely a historical practice but a deeply ingrained conceptual framework [11]. Without a foundational shift in how medical knowledge is constructed and interpreted, AI systems will continue to reflect and amplify this systemic bias [3]. This means AI development must be informed by critical social theory, not just computer science, to avoid encoding existing biases into new technologies [12].

2.1.3 The Epistemological Challenge of the "Male-Default" in Medical AI

The "male-default" is not merely a historical practice but a deeply ingrained conceptual framework that influences how medical knowledge is constructed and interpreted. Without a foundational shift in this understanding, AI will continue to reflect and amplify this systemic bias. This suggests that the problem extends beyond data quantity or technical algorithms; it concerns the very lens through which medical reality is understood and encoded. Consequently, achieving equitable AI demands more than technical fixes; it requires a profound, interdisciplinary engagement with the historical and sociological foundations of medicine. AI developers must collaborate with social scientists, ethicists, and historians to deconstruct and rebuild medical knowledge bases that are truly inclusive, rather than simply digitizing existing, flawed paradigms. This illustrates that ethical AI is fundamentally intertwined with the philosophy of science and medical history, necessitating a re-evaluation of how medical knowledge is generated and disseminated.

2.1.4 The Intersectionality Framework [13] for Comprehensive Equity

Intersectionality is a theoretical framework, initially coined by Kimberlé Williams Crenshaw in 1989, positing that individuals possess multiple social identities—such as race, gender, socioeconomic status, sexual orientation, disability, migration status, and citizenship—that intersect and interact to produce unique experiences of discrimination and marginalization [14]. This framework challenges traditional approaches to understanding health disparities, which often isolate a single factor such as race or socioeconomic status [15]. It highlights how multiple experiential conditions—biological, social, environmental, and economic—converge at individual, community, and structural levels, disproportionately creating barriers and privileges among groups and individuals [15]. This complex interplay can obscure existing inequities if research fails to account for all relevant factors in analyses [15].

Intersectional analysis is crucial for revealing the intricate interactions amongst these social determinants, illustrating the compounded social disadvantages that necessitate targeted policy interventions [15]. For example, an intersectional analysis of chronic disease in Aragon, Spain, showed that low-income migrant women living in urban areas for more than 15 years faced over three times the risk of multimorbidity compared to high-income, non-migrant, urban men, revealing how intersecting disadvantages amplify health risks [16]. Such findings demonstrate the need for nuanced, stratified data in health research and policy.

The intersectionality framework offers precise insights into who is affected and how in different settings and enables policies and programs to identify whom to focus on, whom to protect, what exactly to promote, and why, serving as a scalpel for policies rather than the current hatchet [17]. Therefore, intersectionality is not merely a theoretical lens but a practical, methodological imperative for AI in health equity. Engaging disproportionately impacted priority populations during the development and validation phases ensures the representation of diverse perspectives, while supporting the identification of potential biases before implementation [18].

2.2 The Influence of Social Determinants of Health (SDOH)

Social Determinants of Health (SDOH) are defined as the social conditions in which individuals are born, live, learn, work, play, worship, and age, encompassing factors such as economic stability, education, neighborhood and built environment, health care access, and social context [19]. These factors profoundly influence the distribution of health inequalities and chronic conditions within populations, shaping both risks and outcomes across communities [19].

Research indicates that broader determinants—such as education, employment, food security, and income—can account for up to 50–55% of health outcomes, whereas clinical care typically accounts for only about 20% [20]. Health disparities do not exist in isolation but are part of a reciprocal and complex web of problems associated with inequality and inequity in education, housing, and employment [21]. These disparities disproportionately affect children in racial and ethnic minority, low-income homes and neighborhoods, who experience higher rates of adverse health outcomes and reduced access to resources [22].

The significant impact of SDOH on health outcomes, combined with their frequent exclusion from AI models, creates a substantial blind spot for AI systems [23]. Even if technically robust on clinical data, AI models may produce biased or inaccurate predictions because they lack crucial contextual information that fundamentally drives health outcomes—a phenomenon known as "unmeasured confounding" [23]. If AI systems do not account for factors such as poverty, systemic racism, or lack of healthcare access, their predictions, even if statistically accurate on available data, may be fundamentally flawed or inadvertently perpetuate disparities by misinterpreting correlations as causal clinical factors [23]. This underscores the critical need to integrate non-traditional data sources into AI development and for AI systems to acknowledge their limitations when comprehensive SDOH data are unavailable [24].

3 Algorithmic Bias in AI Disease Prediction: Sources and Consequences

The integration of AI into healthcare offers significant promise but is accompanied by persistent risks of perpetuating and amplifying existing biases throughout the AI lifecycle [3]. These biases can manifest at multiple stages, including data collection, model development, deployment, and real-world application, ultimately influencing clinical decision-making and health equity [3].

3.1 Sources of Biases

3.1.1 Data-Centric Biases [3]

Biases originating from data are among the most critical challenges in AI disease prediction. Imbalanced sample sizes, where certain groups (such as women or racial minorities) are underrepresented, can lead to models that perform poorly for these populations, thereby reinforcing health disparities. Non-randomly missing data, such as incomplete electronic health records for low-income or marginalized groups, further skews model predictions. Additionally, important factors like SDOH—including socioeconomic status, education, and access to care—are often missing or inconsistently captured, resulting in models that overlook key drivers of health outcomes.

Biases in data labeling, such as those introduced by provider cognitive biases or systemic misclassification, can also be embedded in training datasets. This perpetuates historical patterns of discrimination and leads to inequitable AI outputs. The use of race and ethnicity as predictive factors, despite being social rather than biological constructs, has been shown to misrepresent risk and exacerbate inequities when not carefully contextualized.

3.1.2 Algorithmic and Model-Centric Biases

Algorithmic bias can arise during model development when design choices—such as feature selection or optimization metrics—unintentionally prioritize majority groups, leading to differential performance across subpopulations [25]. Overreliance on aggregate performance metrics may obscure poor outcomes for minority groups, while lack of model interpretability (the "black box" problem) impedes identification and correction of biased outputs [3]. Subgroup analysis and bias-centered optimization metrics are essential for detecting and mitigating these issues [3].

3.1.3 Human and Systemic Biases [26]

Human biases, whether implicit or systemic, can enter the AI pipeline through subjective decisions in data labeling, annotation, or model validation. Lack of diversity in AI development teams may introduce blind spots, while broader institutional norms and policies can reinforce inequities at the structural level. Systemic bias is particularly problematic, as it reflects entrenched societal patterns that are difficult to address through technical solutions alone.

3.1.4 Deployment Biases

Even well-designed AI systems may encounter deployment bias when introduced into real-world clinical environments that differ from the original training context [27]. This often results from mismatches between model assumptions and real-world application environments [27]. Publication bias, where studies with positive results or certain medical domains are overrepresented, further distorts the landscape of AI research and its practical applications [3]. These factors contribute to a self-reinforcing cycle, where biased data and models perpetuate and magnify health inequities [25].

The interaction among biased data, subjective human judgment, and systemic inequality creates a self-reinforcing cycle where algorithmic bias replicates and magnifies existing health inequities (Fig. 1).

3.2 The Ethical Dilemma: Balancing False Positives and False Negatives

AI systems in disease prediction inevitably produce errors characterized into two types, both false positives and false negatives, presenting a critical ethical dilemma in high-stakes applications [3].

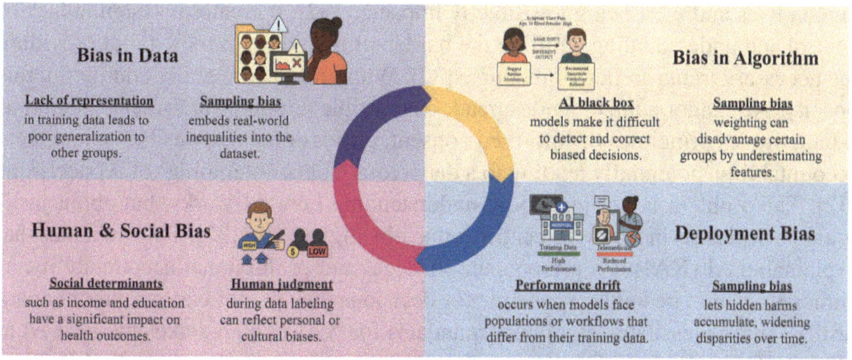

Fig. 1 Illustrating the interactions among biases

False Negatives occur when an AI system incorrectly indicates the absence of a condition when it is actually present. In medical diagnosis, a false negative means missing a true disease, which can delay crucial treatment and potentially lead to severe health outcomes or even death [28].

False Positives occur when an AI system incorrectly indicates the presence of a condition when it objectively does not. In medical diagnosis, such as in cancer screening, a false positive can cause unnecessary anxiety for the patient, lead to further invasive diagnostic procedures, or result in unnecessary treatments for a healthy individual [28].

A fundamental trade-off exists between these error types: reducing one often increases the other. The choice of which error to prioritize depends heavily on the specific consequences of each error type in a given application [28]. For instance, in cancer screening, it might be ethically preferable to accept more false positives to minimize false negatives, given the severe consequences of a missed diagnosis. This highlights that "accuracy" is not a monolithic concept in healthcare AI; its ethical implications are highly context-dependent [29]. This means AI design and evaluation must shift from a universal "best accuracy" approach to a nuanced, application-specific "ethically optimal error balance," requiring close collaboration between AI developers, clinicians, and ethicists to define acceptable risk profiles for different diseases and patient populations [3].

Compounding this dilemma is the "black box" nature of many AI models, particularly deep learning systems, due to their opaque internal decision-making processes [30]. This opacity erodes public trust in AI systems, complicates the assignment of responsibility and accountability when errors occur, and hinders the ability of developers to debug and optimize models [31]. In healthcare, this lack of transparency prevents patients from fully understanding their diagnoses and actively participating in shared decision-making processes [32]. The 'black box' is not merely a technical challenge of interpretability [33]; it is a fundamental barrier to the ethical and responsible deployment of AI, especially in high-stakes domains like healthcare where human lives and well-being are directly impacted [32]. It creates a systemic lack of control and understanding, shifting the burden of trust onto users without providing the necessary transparency or recourse [32]. Without fundamentally addressing this opacity, AI cannot achieve widespread, responsible adoption in critical sectors, as it undermines principles of informed consent, professional responsibility, and legal accountability, potentially leading to a dangerous 'rubber-stamping' of AI decisions [32]. The problem is not just about understanding how AI works, but about justifying its decisions in a human-centric, ethically aligned way [32]. This indicates that explainable AI (XAI) is a prerequisite for legal and professional accountability, as clinicians cannot be held responsible for decisions they do not understand or cannot justify. This necessitates regulatory mandates for XAI in high-stakes medical AI to ensure informed consent and prevent the blind acceptance of potentially biased or erroneous AI outputs [34].

Table 1 provides a concise overview of these error types and their implications:

This table clarifies the concepts of false positives and false negatives by providing clear definitions and concrete clinical examples, such as a cancerous region being overlooked or a healthy region being flagged as cancerous. By explicitly listing the "Consequences for Patient," including "Delayed diagnosis, worsened medical outcomes, increased mortality risk" or "Unnecessary anxiety, additional tests, invasive procedures, financial burden," the table elevates the discussion beyond purely technical metrics to highlight the real-world, human stakes involved in AI diagnostic errors. This reinforces the ethical imperative of the chapter. Furthermore, the inclusion of "Relevant AI Metrics" (recall/sensitivity for false negatives, precision for false positives) provides a direct, actionable link between the ethical consequences and the technical choices in AI development. It demonstrates that the selection of which metric to optimize is not merely a technical decision but a profound ethical one, directly influencing patient outcomes.

Table 1 Typology of AI diagnostic errors and clinical consequences

Error type	Definition	Clinical example	Consequences for patient	Relevant AI metrics
False negative	Fails to detect a condition that is present	Cancerous region overlooked, leading to delayed diagnosis	Delayed diagnosis, worsened medical outcomes, increased mortality risk	Recall/sensitivity (focuses on capturing all actual positives)
False positive	Incorrectly indicates presence of condition	Healthy region flagged as cancerous in medical imaging	Unnecessary anxiety, additional tests, invasive procedures, financial burden	Precision (focuses on reliability of positive predictions)

4 Empirical Evidence of Diagnostic Disparities in AI

Theoretical concerns regarding AI bias are increasingly substantiated by empirical evidence, particularly through real-world instances of disproportionate diagnostic errors among diverse demographic groups. These disparities underscore the necessity of integrating sex, gender, and intersectional analyzes into the design and development of AI systems in healthcare [10].

4.1 Gender-Specific Disparities

AI-driven diagnostic tools have demonstrated significant limitations in accurately diagnosing conditions in women, often leading to critical false negatives or inappropriate false positives.

4.1.1 False Negatives (Missed Diagnosis)

Cardiovascular Disease [35]: Early AI-based cardiovascular risk models consistently underestimated risk in women. Trained predominantly on male-centric datasets, these systems failed to recognize symptoms more common in women—such as nausea, fatigue, and jaw pain—contributing to persistent patterns of misdiagnosis and dismissal in clinical encounters. While the cited source discusses the critical need for AI in women's CVD screening and notes the underrepresentation of women in cardiology research, it also highlights that "female-specific factors have unfortunately not been incorporated into any cardiovascular risk assessment tool", implying a systemic gap that affects AI models. Broader literature further supports that AI models trained on male-dominated datasets can significantly underestimate risk in women.

Liver Disease [36]: Research from University College London reported that AI models predicting liver disease from blood tests missed 44% of cases in women, compared to 23% in men. This disparity stemmed from training data skewed towards male patients and the use of biomarkers calibrated to male physiology.

Chronic and Underdiagnosed Conditions [10]: AI further contributes to the chronic underdiagnosis of conditions primarily affecting women, such as endometriosis, PCOS, autoimmune disorders, and chronic pain syndromes. These limitations arise from the absence of representative data reflecting the full range of female physiological and symptomatic presentations.

4.1.2 False Positives (Incorrect Diagnosis)

Psychiatric Diagnosis [37]: Natural Language Processing (NLP) models used in psychiatric diagnostics have exhibited gender biases, often reflecting and amplifying existing societal stereotypes. While the cited source focuses on the importance of including diverse voices, particularly LGBTQ + communities, to mitigate general biases in AI for mental health, it does not specifically detail the misclassification pattern of men with trauma symptoms being diagnosed with post-traumatic stress disorder (PTSD) while women with similar symptoms are disproportionately diagnosed with personality disorders. This misclassification, whether human or AI-driven, can result in inappropriate treatment and enduring stigma.

The pattern of AI reflecting and amplifying existing human biases, particularly the tendency to dismiss women's physical symptoms as psychological or to misdiagnose psychiatric conditions based on gender stereotypes [10], is a critical feedback loop. This means AI is not merely a neutral tool processing data, but an active participant in perpetuating systemic diagnostic inequities.

4.2 *Racial Disparities*

AI models in healthcare have demonstrated racial bias, particularly manifesting in higher false negative rates for marginalized racial groups. These disparities are often rooted in the structural inequities embedded within clinical data systems [38].

Kidney Disease [38]: A widely used equation for estimating glomerular filtration rate (eGFR) included a "race correction" that overestimated kidney function in Black patients by up to 16%. This led to delayed diagnoses and reduced eligibility for interventions such as dialysis and transplantation.

Skin Cancer [39]: AI models trained primarily on images of lighter skin tones have demonstrated poor performance in detecting melanoma in individuals with darker skin. These tools often fail to recognise atypical lesion locations more common among darker-skinned patients (e.g., palms, soles), due to underrepresentation in training data.

Sepsis and General Illness [40]: Black patients are statistically less likely to undergo diagnostic testing for severe conditions such as sepsis. As a result, training data underrepresents illness severity in these populations, leading AI systems to underestimate clinical urgency.

Occult Hypoxemia [41]: Black patients are three times more likely than white patients to experience undetected low blood oxygen levels (occult hypoxemia) due to inaccuracies in pulse oximetry readings.

Resource Allocation [42]: An algorithm designed to allocate healthcare resources systematically underestimated the health needs of Black patients relative to white patients with similar clinical profiles. This was due to the algorithm's use of historical healthcare expenditure as a proxy for need—an approach that overlooked the impact of systemic underinvestment in Black patients. This racial bias significantly reduced the number of Black patients identified for additional care. The "Resource Allocation" example vividly illustrates how AI can automate and scale structural racism within healthcare. By using a proxy (cost) that is itself a product of historical and systemic inequities (underinvestment in Black communities), the AI system, despite being "accurate" in its cost prediction, perpetuates and amplifies discriminatory practices, transforming historical injustice into algorithmic injustice. This highlights that AI can perpetuate structural racism if not carefully designed, becoming a tool for automating and scaling existing injustices, necessitating a critical examination of the proxies AI models use and their potential to embed and amplify societal biases.

4.3 Intersectional Disparities [43]

Intersectional bias refers to the compounded disadvantages experienced by individuals who embody multiple marginalized identities—such as race, sex, age, or gender diversity. These overlapping dimensions of exclusion intensify disparities in AI-driven healthcare, as demonstrated by empirical studies showing that intersectional subgroups consistently receive lower-quality diagnoses and face greater discrepancies in health outcomes.

Specific Intersectional Cases:

Women of Color in Image Recognition [43]: Image recognition systems have struggled to identify women of color accurately, revealing compounded gender and racial bias in training data and algorithmic design.

Maternal Health for Black and Hispanic Women [44]: AI models predicting the likelihood of successful vaginal birth after Caesarean section (VBAC) have exhibited bias against Black/African American and Hispanic/Latino women. These calculators often included a 'race correction' factor that penalized these groups, making them less likely to be recommended for a trial of labor after cesarean (TOLAC) despite clinical parity with white patients. This bias, rooted in the normalization of race as a biological variable rather than a social construct, contributed to higher rates of unnecessary Caesarean deliveries among these groups.

Elderly Patients and Digital Exclusion [45]: Older adults often face barriers related to digital literacy and age-related bias in technology design. The cited source highlights how implicit age biases can be embedded in AI systems for aged care, leading to paternalistic AI.50 While this source focuses on ageism, the intersectionality framework posits that disadvantages can compound when individuals embody multiple marginalized identities.1 Therefore, if an elderly patient is also from a racial minority group and female, these overlapping disadvantages can indeed intensify, reducing the accessibility and efficacy of AI-based healthcare solutions.

The empirical examples of intersectional bias demonstrate that biases are not merely additive but multiplicative. This means that individuals at the intersection of multiple marginalized identities face disproportionately amplified disadvantages in AI-driven healthcare. This reinforces the concept of intersectionality as a "scalpel" for precision health equity, highlighting the inadequacy of single-axis fairness metrics and necessitating a shift towards more sophisticated, multi-dimensional fairness metrics and data collection strategies that capture these complex interactions [43]. Table 2 summarizes illustrative cases of AI bias.

This table serves as a powerful summary of the real-world impact of AI bias. By explicitly linking error types, demographic groups, diseases, specific manifestations of error, and their underlying causes, it clarifies the causal chain from data and systemic issues to patient harm. This is crucial for an academic report as it moves beyond abstract discussions to demonstrate the tangible consequences of unaddressed biases, making the case for mitigation strategies more compelling.

5 Case Study: AI Models in Cardiovascular Disease—Bias and Mitigation

Historically, conditions such as CVD, osteoporosis, and chronic pain have demonstrated notable gender disparities in both diagnosis and treatment [46]. This section focuses on CVD, highlighting how AI models can either perpetuate existing gender biases or enhance diagnostic equity when designed with sex- and gender-sensitive methodologies.

5.1 AI Applications and Gender Bias in Cardiovascular Disease

CVD has traditionally been conceptualized as a male-dominant condition, despite it being the leading cause of death among women globally. Clinical diagnostics and treatment protocols have predominantly been developed around male pathophysiology, resulting in significant knowledge gaps and poorer outcomes for women [25]. Emerging research shows that CVD in women—particularly younger women—often presents through distinct biological mechanisms, such as non-obstructive coronary conditions like microvascular dysfunction, which are frequently undetected by conventional diagnostic tools such as angiography [25].

Women often present with symptoms such as nausea, fatigue, and jaw pain—symptomatology that deviates from the typical male symptom profile, increasing the likelihood of misdiagnosis [47]. Gender-related factors, including social norms, behaviors, and roles, have been shown to independently affect prognosis. Notably,

Table 2 Illustrative cases of AI bias in healthcare by demographic and error type

Error type	Demographic group	Disease/ condition	Specific manifestation of error	Underlying cause of bias
False negative	Women	Cardiovascular disease (CVD)	Failure to detect female-specific symptoms (e.g. nausea, fatigue, jaw pain) due to male-centered training data	Predominantly male training data; insufficient representation of diverse symptomatology
	Women	Liver disease	AI models missed 44% of female cases compared to 23% of male cases	Over-reliance on male-centric biomarkers and imbalanced training datasets
	Black	Kidney disease	Delayed diagnosis due to overestimated kidney function by race-based eGFR formula	Use of flawed race correction in algorithm, grounded in discredited biological assumptions
	Black	Skin cancer	Inability to accurately identify lesions on darker skin tones, particularly in atypical locations	Underrepresentation of darker skin images in training datasets
	Black	Sepsis/general illness	Underestimated illness due to lower rates of medical testing in training data	Systemic undertesting encoded into training data
	Black	Resource allocation	Underestimated health needs based on lower historical health expenditure	Use of healthcare expenditure as a proxy for need, reflecting systemic underinvestment

(continued)

Table 2 (continued)

Error type	Demographic group	Disease/ condition	Specific manifestation of error	Underlying cause of bias
False positive (mischaracterization)	Women	Psychiatric diagnosis	Trauma symptoms in women more likely to be misdiagnosed as personality disorders instead of PTSD	AI replication of gender stereotypes in natural language processing and diagnostic frameworks
False positive (unnecessary C-section)	Black/ Hispanic women	Maternal health (VBAC)	Higher rates of C-sections compared to White women despite similar conditions	Developer assumptions and historical biases embedded in algorithms

PTSD post-traumatic stress disorder, *VBAC* vaginal birth after Caesarean section, *eGFR* estimating glomerular filtration rate

individuals of any sex who express more traditionally "feminine" traits have demonstrated poorer outcomes following cardiac events, indicating that gender expression can influence health behaviors and clinical interactions [48]. These considerations reveal how historical male bias in cardiovascular datasets has led to the development of AI systems that reinforce diagnostic disparities.

5.2 AI Models Exhibiting Gender Bias in Cardiovascular Disease

Several studies have documented how AI models used in CVD diagnostics can perpetuate gender disparities when developed without disaggregated data or consideration of sex-specific variables.

5.2.1 Babylon Health's AI-Powered Symptom Checker [49]

In a reported case, a 59-year-old female smoker presenting with chest pain and nausea was advised that her symptoms might indicate anxiety or depression. Conversely, a male patient with an identical profile was instructed to seek emergency care. This divergence was attributed to historical underrepresentation and misdiagnosis of women in the datasets on which the model was trained, thereby embedding those biases into the algorithm.

5.2.2 Large Language Models (LLMs) such as GPT-4 [47]

A 2024 study assessing GPT-4's evaluation of coronary artery disease risk found inconsistencies across gender. While women were initially assigned higher risk scores (citing age as a key factor), the introduction of psychiatric comorbidities shifted the model's interpretation—assigning higher risk to male patients and downgrading women. This pattern illustrates a wider bias whereby women's symptoms are more readily attributed to psychosocial causes, thus potentially undermining clinical accuracy.

5.2.3 AI-Based Medical Imaging [47, 50]

Diagnostic models trained on male-centric or unrepresentative data have exhibited reduced accuracy for female patients and people of color. Studies reveal that these models can perform unevenly across demographics, leading to "fairness gaps". Reports have raised serious concerns about these tools' embedded gender and racial biases, warning that inadequate validation poses risks to clinical safety and equity [34].

The consistent replication of human biases (e.g., dismissing women's symptoms as psychological, male-centric medical imaging) in AI models highlights that AI is not an objective arbiter but a mirror and amplifier of existing societal and clinical prejudices. This means that technical solutions alone are insufficient; a fundamental re-evaluation of medical understanding and data collection practices is required to achieve equitable AI. We summarize the baises in Table 3.

This table is crucial because it moves from general statements about bias to concrete examples. By listing specific models and their documented biases and impacts, it grounds the theoretical discussion in real-world applications. This strengthens the argument that AI bias is a present and pressing concern, not a hypothetical one, and sets the stage for discussing mitigation strategies.

Table 3 AI models exhibiting gender bias in CVD

Model/platform	Nature of gender bias	Impact
Babylon symptom checker	Under-triages women with cardiac symptoms	Delayed or missed diagnosis
GPT-4 (LLM)	Inconsistent risk predictions linked to gender and psychiatric history	Inconsistent risk predictions
Medical imaging AI	Imbalanced datasets	Reduced diagnostic accuracy for women

5.3 *Fairer AI Models: Integrating Sex and Gender Analysis in Cardiovascular Disease*

In contrast to earlier examples of gender-biased AI systems, recent innovations have demonstrated how sex- and gender-sensitive design principles can yield AI models that are both more equitable and clinically effective in the context of CVD [51].

5.3.1 Sex-Specific Cardiovascular Risk Prediction Models

A 2025 study [51] employed an extreme gradient boosting (XGBoost) algorithm to analyze data from over 32,000 individuals, developing separate predictive models and intervention thresholds for men and women. For male patients, the AI model significantly outperformed standard risk assessment tools—such as NORRISK 2 and SCORE2—in terms of both ROC-AUC and precision-recall metrics. While performance improvements for female patients were more modest, the model successfully identified novel, sex-specific risk predictors and introduced age-adjusted intervention thresholds.

5.3.2 AI-Enhanced Electrocardiogram (ECG) Analysis

Researchers at Imperial College London have developed an AI model that can identify women at a higher risk of CVD. This model, trained on over one million ECG records, generates a composite score indicating whether a patient's ECG pattern resembles a typical male or female [52]. A key finding was that women whose ECGs more closely matched a "typical male" pattern—for instance, showing an increased electrical signal—tended to have larger heart chambers and more muscle mass. Crucially, these women also faced a significantly higher risk of cardiovascular disease, future heart failure, and heart attacks compared to women with ECGs more closely matching the "typical female" pattern [53]. This model exemplifies a physiologically informed and sex-aware approach to risk assessment, enabling earlier and more accurate identification of high-risk women.

Complementary evidence comes from a 2023 study [54] using a self-attention deep learning framework to detect left ventricular hypertrophy (LVH) from ECG signals, which also highlighted significant performance differences between sexes (ROC-AUC: 0.826 for men vs. 0.772 for women). Importantly, the class activation map of that study visualized how the model's attention focused differently across the QRS complex depending on sex and LVH status—suggesting that ECG morphologies in women may be less distinguishable by AI, potentially explaining their lower predictive performance. This supports the notion that physiologically and sex-informed modeling is critical for fair and effective cardiovascular risk stratification, addressing the dual challenge of underdiagnosis in women and overtreatment in men.

5.3.3 Rethinking AI-Predicted Sex from ECG: Novel Biomarker

Recent studies have consistently shown that AI models trained on ECG signals can classify a person's sex with remarkable accuracy [55]. This has raised questions about the nature of ECG-encoded sex differences and suggests that sex-specific features may already be deeply embedded in the waveform itself, allowing AI models to implicitly learn them during training. A recent 2025 study addressed this question by introducing the concept of a sex discordance score—the discrepancy between AI-inferred sex and self-reported sex—as a potential new biomarker for cardiovascular risk stratification [52]. Using over one million ECGs from the BIDMC and UK Biobank datasets, the authors found that women with "male-like" ECG patterns (i.e., higher sex discordance scores) were at significantly increased risk of cardiovascular events such as heart failure and myocardial infarction, despite having normal ECGs by traditional clinical standards. In contrast, this association was not observed in men. These findings challenge the notion that AI-predicted sex is merely a curiosity or technical side effect. Instead, they suggest that the mismatch itself may serve as a novel, physiologically grounded biomarker, especially useful for identifying high-risk women who might otherwise be overlooked by conventional assessments.

5.3.4 Multi-Modal Ensemble Model [56, 57]

This integrates diverse data sources including electronic health records, imaging, and wearable devices, offer even greater diagnostic potential. One example involves the repurposing of breast arterial calcification detected during routine mammography as an indicator of cardiovascular risk. These integrative tools have been shown to detect non-traditional risk factors that disproportionately affect women, thereby supporting personalized and earlier intervention.

5.3.5 GRACE (Global Registry of Acute Coronary Events) Risk Score [58]

This score was also updated in 2022 to incorporate AI-based predictive models accounting for sex-specific disease characteristics. This revision (GRACE 3.0) improves performance in both sexes and reduces diagnostic disparities in acute coronary syndromes.

The success of "fairer AI models" in CVD (e.g., sex-specific XGBoost, AI-ECG model, sex discordance score) demonstrates a crucial shift from reactive "debiasing" to proactive "equity-by-design". This implies that AI's impact on equity is not inherent but a direct consequence of intentional design choices, transforming it from a bias amplifier to a tool for health equity. The examples of separate models, sex-specific features, and physiologically informed scores show that equitable AI is achievable through intentional engineering and ethical consideration.

Table 4 AI-aided CVD models integrating sex and gender analysis

Model/study	Sex/gender integration approach	Outcome/impact
XGBoost risk model	Separate models and intervention thresholds by sex	Enhanced precision and identification of female-specific predictors
AI-ECG model (imperial)	Sex-specific ECG pattern analysis	Improved risk detection for high-risk women
Multi-modal ensemble models	Inclusion of imaging, wearables, and EHR data	Broader screening, early detection of female-specific risks
GRACE 3.0 risk score	Sex-specific AI-based update	Improved outcomes and reduced disparities

EHR electronic health records

We summarize the fairer AI-models by integrating sex and gender analysis in Table 4.

These examples demonstrate a shift from reactive debiasing to proactive "equity-by-design" [59]. Instead of fixing bias after it occurs, these models are built with sex and gender as fundamental variables from the outset, using approaches such as separate models, sex-specific features, and physiologically informed scores [59]. This indicates that intentional, informed design can transform AI from a bias amplifier into a tool for achieving health equity. AI's impact on equity is not inherent; it is a direct consequence of its design. This highlights the importance of "gendered innovations" and similar frameworks that integrate sex and gender analysis throughout the research and development lifecycle, demonstrating that equitable AI is achievable through intentional engineering and ethical consideration.

5.4 *Lessons Learned and Future Directions*

The review of AI applications in cardiovascular diagnostics reveals both the pitfalls of perpetuating gender bias and the opportunities for advancing equity through more inclusive design. The evidence underscores the necessity of embedding sex- and gender-informed considerations throughout the lifecycle of AI development—from data collection to clinical deployment.

Key lessons and best practices emerging from this review include:

Holistic and Life-Course Data Collection [25, 60, 61]: Effective AI models must incorporate clinical data reflective of women's health across the life span. Risk factors and physiological changes vary significantly across key stages—such as adolescence, pregnancy, menopause, and older age. Without longitudinal and sex-specific data, AI systems risk missing early predictors of CVD in women.

Multimodal and Wearable Data Integration [57]: The use of multimodal data—ranging from electronic health records and imaging to wearable and remote sensing technologies—offers unprecedented opportunities for comprehensive risk assessment. These data sources provide continuous, real-time physiological inputs that enhance the early detection of atypical or non-traditional risk factors, particularly in women.

Sex- and Gender-Specific Screening Protocols [61]: Screening frameworks must move beyond generalized algorithms to include sex- and gender-informed clinical indicators. Female-specific risk factors—such as pregnancy-related complications, early menopause, and breast arterial calcification—must be systematically incorporated into AI models to improve diagnostic accuracy and ensure earlier intervention.

Balanced and Stratified Data Representation [25, 47]: Equitable representation of both sexes in training datasets is critical. Underrepresentation of women in historical datasets has contributed to systemic diagnostic errors. Sex-stratified modelling enhances algorithmic precision and ensures generalizability across diverse patient populations.

Tailored Feature Selection [51]: The inclusion of variables that reflect physiological and behavioral differences between men and women is vital. Tailored feature selection enables models to detect divergent symptom presentations and risk profiles, reducing both false positives and false negatives.

Clinical Integration and Provider Education [25]: For AI systems to succeed in practice, they must be embedded within clinical workflows and supported by provider training. Clinicians must be empowered to interpret sex- and gender-specific outputs with confidence. Equally, policymakers must facilitate the adoption of tools that align with equity-focused health strategies.

Future Directions

The best practices outlined in this chapter chart a pathway toward a future in which AI actively supports a more just, inclusive, and clinically effective model of cardiovascular care. As the field progresses, several key priorities emerge. It is crucial to develop globally representative, sex-disaggregated datasets to ensure AI models are robust and equitable across diverse populations. Furthermore, embedding intersectional frameworks that account for race, age, gender identity, and SDOH will be vital for addressing health disparities. Advancing XAI is also paramount to improve transparency and trust in clinical decision-making. Finally, strengthening interdisciplinary collaboration among clinicians, data scientists, ethicists, and patient advocates will foster a holistic approach to AI development and implementation.

While AI holds immense potential to transform cardiovascular diagnostics, this transformation will only be fully realized if future systems are built with intentionality, prioritizing fairness, inclusivity, and context-specific precision.

6 Mitigation Strategies: A Multi-Layered Framework for Equitable AI

Effectively addressing bias in AI-powered disease prediction requires a comprehensive, multi-layered strategy that spans data quality, algorithmic design, human oversight, and regulatory governance [3]. This approach must be attuned to the ways in which false positives and false negatives disproportionately affect specific demographic groups and intersectional identities, with the ultimate goal of promoting equity, safety, and clinical reliability [3]. The framework presented herein establishes a systematic methodology for developing AI systems that actively counteract bias rather than perpetuating existing healthcare disparities [62].

6.1 Data-Centric Interventions

The foundation of equitable AI lies in the quality, completeness and representativeness of its training data [3]. Historical datasets in healthcare have systematically underrepresented women, ethnic minorities, and other marginalized populations, leading to AI systems that perform poorly for these groups.

6.1.1 Diverse and Representative Data Collection [3]

It is crucial to ensure that training data encompasses diverse populations across all ages, genders, ethnicities, and socioeconomic backgrounds. This includes actively selecting data that reflects different social backgrounds, cultures, and roles, while systematically removing historical biases that associate specific jobs or traits with one gender. For healthcare, this means prioritizing sex-disaggregated data collection, which breaks down information by biological sex to reveal hidden patterns and inequities in health outcomes. Additionally, techniques such as data augmentation and transfer learning can help overcome data scarcity for underrepresented groups by enhancing sample diversity.

6.1.2 High-Quality Data Annotation [3]

Reliable labelling of medical data is critical to the development of trustworthy AI models. This requires rigorous human validation, even in automated or semi-automated settings. Techniques such as dual annotation—where two independent reviewers label the same data—can improve inter-annotator reliability. Clear annotation guidelines and ongoing training are essential to reduce subjective interpretation and implicit bias.

6.1.3 Data Preprocessing Techniques [3]

Before model training, various preprocessing methods can be applied to mitigate bias. These include relabeling and reweighing data to balance datasets and ensure adequate representation of each group. Moreover, NLP tools can extract contextual information from clinical notes, revealing nuanced patterns that may not be evident in structured data. Early-stage bias mitigation is most effective when supported by open-source data-sharing initiatives and inclusive stakeholder engagement.

6.2 Algorithmic and Model-Centric Interventions

In addition to data-level improvements, the algorithms themselves must be designed and evaluated with fairness as a core principle [63].

6.2.1 Fairness-Aware Algorithms [63]

Model training should incorporate fairness constraints and performance metrics that explicitly measure equity. Examples include "equalized odds", which aims for parity in true and false positive rates across groups, and "equal opportunity", which ensures that true positive rates are consistent among protected categories. Such metrics guide the optimization of models beyond aggregate accuracy. To ensure these fairness objectives are being met, visualizing group-specific performance metrics—such as F1-score or sensitivity across subgroups (e.g., skin tone or sex)—is essential [3]. This enables the detection and communication of performance disparities that may otherwise be hidden when only overall accuracy is reported [3]. In this context, Fairlearn [64], an open-source toolkit developed by Microsoft, facilitates this process by providing algorithms and dashboards designed to assess and mitigate fairness-related problems. Specifically, through both post-processing and in-processing (reduction-based) methods, Fairlearn allows model developers to balance performance and equity across subgroups. Moreover, its interactive visualization tools support side-by-side comparisons of metrics such as accuracy, recall, and error rates by sex or skin tone, thereby helping to identify allocation or service quality harms and guide more equitable model selection. Similar fairness-aware approaches have also been applied to generative models. FairDiffusion, for example, incorporates fairness constraints into diffusion-based medical image generation, ensuring that images produced across subgroups (e.g., sex, race, or skin tone) maintain equitable quality and clinical relevance [65].

6.2.2 Explainable AI (XAI)

The "black box" nature of many AI systems raises critical concerns around trust, accountability, and clinical safety. Explainability tools such as SHAP (Shapley Additive Explanations) and LIME (Local Interpretable Model-agnostic Explanations) provide transparent, interpretable insights into how models generate predictions [66]. This empowers clinicians to make informed decisions and critically evaluate the model's outputs, particularly when related to protected characteristics. Recent study has gone beyond traditional XAI tools by employing clustering analyses, counterfactual image generation, and a novel 'removal via balancing' technique to identify and quantify specific visual signals—such as hair, age-related features, or imaging artifacts—that AI models exploit to predict protected attributes like patient sex from dermoscopic images [67]. These approaches demonstrate the potential of XAI not only to enhance interpretability but also to uncover hidden biases and support fairness in clinical AI systems.

6.2.3 Causal Inference for Fairness [62]

Moving beyond mere correlations, incorporating causal inference into AI development allows for a deeper understanding of the underlying causal relationships between interventions and health outcomes. This approach helps identify the root causes of disparities, rather than just their symptoms, enabling the design of interventions and policies that target these fundamental issues to promote fairness and equity in healthcare delivery. Unlike conventional correlation-based methods, causal inference approaches address the fundamental question: would the clinical decision support system make a different decision if the patient had a different sensitive attribute such as race or gender? Advanced causal-model-free algorithms such as CFReg have been developed to provide causal fairness for any supervised machine learning model whilst maintaining good trade-offs between fairness and classification performance.

6.2.4 Robustness and Generalizability [3]

AI systems should be tested on independent datasets and across different clinical settings to ensure they perform well beyond their original training environment. This is essential to avoid failures when the model is used in real-world scenarios that may differ from the data it was trained on [29]. Rigorous validation and careful consideration of how the model might behave in new or varied contexts are key to ensuring reliability.

6.3 Human-Collaboration and Oversight Mechanisms

Human judgment remains indispensable in the development, deployment, and monitoring of AI systems in healthcare [68].

6.3.1 Continuous Monitoring and Auditing

AI systems require post-deployment surveillance to ensure sustained fairness and clinical efficacy. Regular audits should assess model performance across diverse demographic groups and track whether disparities emerge or intensify over time [3]. This iterative feedback loop is essential for ongoing bias correction and ethical compliance [3]. Tools such as FairLens [69], Aequitas [70], MONET [71], and IBM AI Fairness 360 [72] provide structured auditing workflows for detecting and mitigating demographic-specific disparities. Integrating these into post-deployment pipelines can enhance ethical oversight and transparency.

6.3.2 Human Oversight and Interdisciplinary Collaboration

AI systems in healthcare must function as augmentative tools that enhance rather than replace clinical expertise, ensuring that human judgment remains central to patient care decisions [73]. Healthcare professionals should maintain ultimate decision-making authority and possess the capability to override AI-generated recommendations when clinical circumstances warrant such intervention [74]. The creation and implementation of healthcare AI systems necessitates collaborative involvement from diverse professional stakeholders, including clinicians, data scientists, ethicists, sociologists, and representatives from marginalized communities. Research demonstrates that multidisciplinary teams are essential for identifying potential biases, addressing ethical considerations, and ensuring that AI systems serve all demographic groups equitably [68].

6.3.3 Patient and Public Involvement [68]

Actively involving marginalized communities and patients in the design and evaluation of AI systems is crucial to ensure their needs and perspectives are reflected. This participatory approach fosters equity and inclusivity, building trust and ensuring that AI solutions are truly patient-centered. Patients should be informed about AI use in their care and have the right to consent or opt-out.

6.4 *Policy, Regulatory, and Ethical Governance*

To achieve equitable outcomes with AI, robust policy frameworks and regulatory oversight are crucial.

6.4.1 Regulatory Oversight and Standards [75, 76]

Effective regulatory oversight and standards are crucial for health AI tools, especially those deemed "high-risk" by authorities. Both European and American regulatory bodies have established frameworks emphasizing transparency, validation, and monitoring. The European Artificial Intelligence Act, effective August 1, 2024, represents a pioneering comprehensive AI regulatory framework. It classifies medical AI devices as high-risk, necessitating rigorous risk management, EU AI database registration, conformity assessment under both the AI Act and Medical Device Regulation, high-quality representative datasets, and robust human oversight mechanisms, with full compliance anticipated by 2027. Similarly, the U.S. Department of Health and Human Services finalized the "Health Data, Technology, and Interoperability (HTI-1)" rule in December 2023, introducing vital transparency requirements for AI in certified health IT systems, impacting the majority of U.S. healthcare providers. Both frameworks mandate comprehensive disclosures from developers regarding software methodologies, funding, clinical applications, training data characteristics, performance across demographic subgroups, and ongoing monitoring. These requirements collectively aim to enhance fairness, patient safety, and clinician trust in AI by ensuring thorough validation, post-market surveillance, and the disclosure of subgroup-specific performance metrics in approval applications.

6.4.2 Ethical Principles [3]

AI in healthcare must adhere to the core ethical principles of biomedical practice—autonomy, beneficence, non-maleficence, and justice [8]. These principles should be embedded from the problem identification stage through data gathering, algorithm development, and model implementation. The integration of ethical considerations throughout the AI development lifecycle ensures that systems serve the interests of all patients equitably.

6.4.3 Transparency and Accountability [3]

Regulatory frameworks should mandate transparent reporting on AI performance across subgroups and clarify liability for errors [5]. Developers should provide detailed documentation of model assumptions, data sources, demographic coverage, and subgroup performance. Where errors occur, liability and responsibility must be clearly delineated. Adopting open science principles—such as sharing source code and metadata—can further strengthen trust and facilitate peer and regulatory review.

6.4.4 Promoting AI Literacy [3, 77]

Education and training are vital for all stakeholders. Healthcare professionals must understand when to apply specific healthcare AI, the level of confidence to place in algorithmic recommendations, and how to evaluate a model's performance. This includes recognizing the limitations of AI systems and understanding when human oversight is particularly critical. Patients also need basic AI literacy to understand its benefits, risks, and how to engage meaningfully with AI-supported care. As AI becomes more prevalent in healthcare settings, patient education programs should include information about how AI systems work, their potential benefits and limitations, and patients' rights regarding AI-assisted care.

The multi-layered framework for mitigation strategies highlights that ethical AI is not a single technical fix but a continuous, systemic endeavor involving technical, social, and governance interventions. The challenge of "misaligned incentives" and "low confidence in AI within a fragmented regulatory and governance framework" reveals that even scientifically sound solutions face significant implementation hurdles. This implies that the success of equitable AI hinges not just on developing the right algorithms or collecting diverse data, but on overcoming systemic barriers like fragmented governance, lack of AI literacy among leaders, and misaligned incentives. This underscores that policy and education are as crucial as technical advancements for real-world impact.

Ultimately, creating fairer and more trustworthy medical AI necessitates interventions across data, algorithmic, and human levels, complemented by robust policy, regulation, and governance, as summarized in Fig. 2.

6.5 Implementation Challenges [78]

Ultimately, scaling AI in healthcare faces significant hurdles, as highlighted by recent research from The World Economic Forum Digital Healthcare Transformation (DHT) initiative. Three major challenges impede its progress: Firstly, the inherent complexity of AI in health deters policymakers and business leaders. Despite considerable attention, AI struggles to gain traction on political and strategic agendas. Secondly, there's a misalignment between technical choices and strategic visions. Health leaders often delegate technical decisions, missing crucial opportunities to align technology with their overarching strategic goals. This is further compounded by misaligned incentives that hinder decisions supporting shared objectives and collective ideals. Finally, low confidence in AI within a fragmented regulatory and governance framework poses a significant barrier. Growing public distrust in AI and industry skepticism could severely impede its widespread adoption in healthcare.

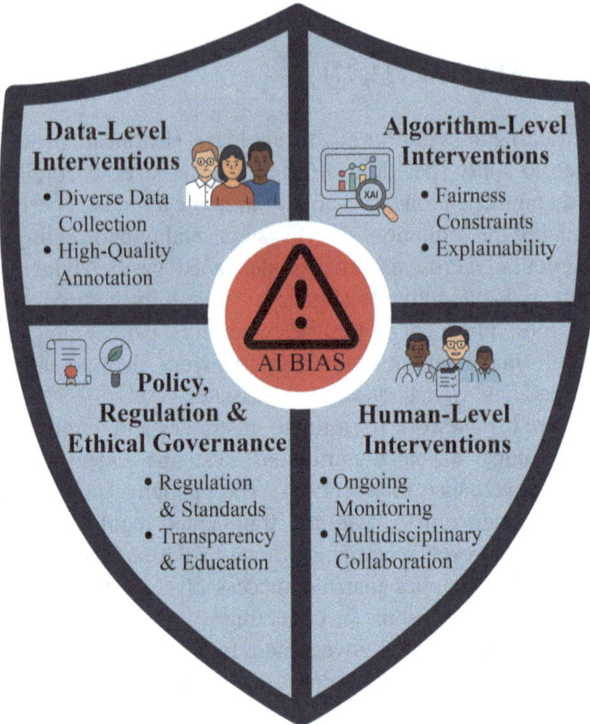

Fig. 2 Toward fair and trustworthy medical AI

7 Conclusion and Future Directions

The Concluding Remarks

Integrating AI into disease prediction presents a transformative opportunity, yet it also poses significant challenges, especially concerning fairness, accuracy, and accountability. As demonstrated throughout this chapter, AI systems aren't inherently neutral; instead, they often reflect and amplify existing structural inequalities embedded in historical medical research and clinical practice.

Diagnostic disparities—particularly false positives and false negatives—are not evenly distributed [3]. They tend to disproportionately affect individuals based on sex, gender, race, age, and their intersections. At the core of this issue lies the persistent "default patient" assumption, which centers AI development on a narrow demographic profile—typically white, male, and able-bodied. This marginalizes those outside this profile, contributing to misdiagnoses, delayed treatments, and poorer health outcomes.

Compounding this is the "black box" nature of many AI models, which hinders transparency and makes it difficult to scrutinize decision-making processes [79]. This opacity weakens trust among clinicians and patients, limits opportunities for meaningful oversight, and complicates efforts to detect or correct embedded biases [3]. As a result, these tools may perpetuate harm under the guise of innovation if not critically and ethically designed.

Achieving health equity through AI necessitates a fundamental paradigm shift: integrating sex and gender analysis at the core of AI development. This extends beyond merely augmenting datasets with more information on women and marginalized populations; it demands a critical re-evaluation of existing medical knowledge and data collection methodologies. The prevalent "male-default" in medical understanding must be challenged through deliberate AI design choices that acknowledge and incorporate sex and gender-based differences in disease presentation, progression, and treatment response. Empirical evidence indicates that incorporating sex and gender as foundational variables from the initial stages of AI development can enhance, rather than diminish, diagnostic equity. Approaches such as developing sex-specific models, integrating sex-informed features, and utilizing physiologically grounded scoring systems have demonstrated promise in mitigating diagnostic disparities. This proactive "equity-by-design" framework transforms AI from a potential amplifier of bias into a powerful instrument for achieving health equity. Initiatives like the MESSAGE framework (2024) [80] represent critical advancements toward institutionalizing sex and gender analysis within health research and AI development.

Beyond the considerations of sex and gender, an intersectional approach is indispensable for cultivating truly equitable AI systems. Intersectionality posits that individuals possess multiple, intersecting social identities—such as race, gender, socioeconomic status, and disability—which coalesce to produce unique experiences of discrimination and marginalization. This framework challenges conventional paradigms that isolate single factors in the analysis of health disparities. Research demonstrates that AI models designed with intersectionality as a core principle can more effectively capture and address the nuanced and compounded effects of multiple, intersecting disadvantages [81].

The Path Forward [3, 26, 82]

Tackling these challenges requires more than technical optimization—it demands a rethinking of how AI is conceptualized, developed, and deployed in healthcare. Ethical, clinical, and social dimensions must be fully integrated into every stage of the AI lifecycle. This includes ensuring that development teams are diverse and interdisciplinary, that models are tested on representative populations, and that patients are informed and empowered participants in AI-supported care.

Effective AI lifecycle management now demands new competencies beyond what traditional tech governance frameworks have required. Ensuring AI systems serve as tools for equitable and ethical care requires targeted interventions and a commitment to both technological progress and social responsibility. This includes developing globally representative, sex-disaggregated datasets, embedding intersectional frameworks that account for race, age, gender identity, and SDOH, advancing explainable AI to improve transparency in clinical decision-making, and strengthening interdisciplinary collaboration among clinicians, data scientists, ethicists, and patient advocates.

Ultimately, for AI to contribute to better, fairer health outcomes, it must be grounded in principles of equity, transparency, and inclusivity. The path forward lies in intentionally dismantling the biases that have historically shaped health data and decision-making—and building AI systems that serve all individuals with equal care and accuracy.

References

1. Dankwa-Mullan I. Health equity and ethical considerations in using artificial intelligence in public health and medicine. Prev Chronic Dis. 2024;21:E64.
2. Thomas LB, Mastorides SM, Viswanadhan NA, Jakey CE, Borkowski AA. Artificial intelligence: review of current and future applications in medicine. Fed Pract. 2021;38:527–38.
3. Cross JL, Choma MA, Onofrey JA. Bias in medical AI: implications for clinical decision-making. PLOS Digit Health. 2024;3:e0000651.
4. Chen RJ, Wang JJ, Williamson DFK, Chen TY, Lipkova J, Lu MY, et al. Algorithmic fairness in artificial intelligence for medicine and healthcare. Nat Biomed Eng. 2023;7:719–42.
5. Canadian Institutes of Health Research. What is gender? What is sex? 2023. https://cihr-irsc.gc.ca/e/48642.html. Accessed 30 Jun 2025.
6. Migliore L, Nicoli V, Stoccoro A. Gender specific differences in disease susceptibility: the role of epigenetics. Biomedicines. 2021;9:652.
7. Sun TY, Hardin J, Nieva HR, Natarajan K, Cheng RF, Ryan P, et al. Large-scale characterization of gender differences in diagnosis prevalence and time to diagnosis. medRxiv 2023. https://doi.org/10.1101/2023.10.12.23296976.
8. Shi Y, Ma J, Li S, Liu C, Liu Y, Chen J, et al. Sex difference in human diseases: mechanistic insights and clinical implications. Signal Transduct Target Ther. 2024;9:238.
9. Mehta LS, Beckie TM, DeVon HA, Grines CL, Krumholz HM, Johnson MN, et al. Acute myocardial infarction in women: a scientific statement from the American Heart Association. Circulation. 2016;133:916–47.
10. Alvaro D. The gender bias built into AI—and its threat to women's health; 2025. https://www.pharmasalmanac.com/articles/the-gender-bias-built-into-ai-and-its-threat-to-womens-health. Accessed 30 Jun 2025.
11. Merone L, Tsey K, Russell D, Nagle C. Sex inequalities in medical research: a systematic scoping review of the literature. Womens Health Rep (New Rochelle). 2022;3:49–59.
12. Buslon N, Cortes A, Catuara-Solarz S, Cirillo D, Rementeria MJ. Raising awareness of sex and gender bias in artificial intelligence and health. Front Glob Womens Health. 2023;4:970312.
13. Stefanick ML, Schiebinger L. Analysing how sex and gender interact. Lancet. 2020;396:1553–4.
14. Public Health Agency of Canada. How to integrate intersectionality theory in quantitative health equity analysis? A rapid review and checklist of promising practices. Ottawa, ON; 2022.
15. Harari L, Lee C. Intersectionality in quantitative health disparities research: a systematic review of challenges and limitations in empirical studies. Soc Sci Med. 2021;277:113876.
16. Moreno-Juste A, Gimeno-Miguel A, Poblador-Plou B, Calderon-Larranaga A, Cano Del Pozo M, Forjaz MJ, et al. Multimorbidity, social determinants and intersectionality in chronic patients. Results from the EpiChron Cohort. J Glob Health. 2023;13:04014.

17. Christoffersen A. Applying intersectionality in policy and practice: unseating the dominance of gender in responding to social inequality. sozialpolitik.ch. 2023;1/2023:1.7.
18. Ghanem S, Moraleja M, Gravesande D, Rooney J. Integrating health equity in artificial intelligence for public health in Canada: a rapid narrative review. Front Public Health. 2025;13:1524616.
19. Healthy People 2030. Social Determinants of Health. https://odphp.health.gov/healthypeople/priority-areas/social-determinants-health. Accessed 30 Jun 2025.
20. Deng IX, Shih P. Social determinants of health: the unaddressed variable accounting for 80% of health outcomes. CareJourney. 2020. https://carejourney.com/social-determinants-of-health/. Accessed 30 June 2025.
21. Canadian Public Health Association. What are the social determinants of health? https://www.cpha.ca/what-are-social-determinants-health. Accessed 30 Jun 2025.
22. Hoffmann JA, Alegría M, Alvarez K, Anosike A, Shah PP, Simon KM, et al. Disparities in pediatric mental and behavioral health conditions. Pediatrics. 2022;150:e2022058227.
23. Guevara M, Chen S, Thomas S, Chaunzwa TL, Franco I, Kann BH, et al. Large language models to identify social determinants of health in electronic health records. Npj Digital Med. 2024;7:6.
24. Abbott EE, Apakama D, Richardson L, Chan LL, Nadkarni GN. Leveraging artificial intelligence and data science for integration of social determinants of health in emergency medicine: scoping review. JMIR Med Inform. 2024;12:e57124.
25. Mihan A, Pandey A, Van Spall HGC. Artificial intelligence bias in the prediction and detection of cardiovascular disease. NPJ Cardiovasc Health. 2024;1:31.
26. Hasanzadeh F, Josephson CB, Waters G, Adedinsewo D, Azizi Z, White JA. Bias recognition and mitigation strategies in artificial intelligence healthcare applications. Npj Digital Med. 2025;8:154.
27. Gichoya JW, Thomas K, Celi LA, Safdar N, Banerjee I, Banja JD, et al. AI pitfalls and what not to do: mitigating bias in AI. Br J Radiol. 2023;96:20230023.
28. Brown SD. Ethical challenges in child abuse: what is the harm of a misdiagnosis? Pediatr Radiol. 2021;51:1070–5.
29. Abujaber AA, Nashwan AJ. Ethical framework for artificial intelligence in healthcare research: a path to integrity. World J Methodol. 2024;14:94071.
30. Chaddad A, Peng J, Xu J, Bouridane A. Survey of explainable AI techniques in healthcare. Sensors (Basel). 2023;23:634.
31. Chan B. Black-box assisted medical decisions: AI power vs. ethical physician care. Med Health Care Philos. 2023;26:285–92.
32. Kiseleva A, Kotzinos D, De Hert P. Transparency of AI in healthcare as a multilayered system of accountabilities: between legal requirements and technical limitations. Front Artif Intell. 2022;5:879603.
33. Lang BH, Nyholm S, Blumenthal-Barby J. Responsibility gaps and black box healthcare AI: shared responsibilization as a solution. Digit Soc. 2023;2:52.
34. Palaniappan K, Lin EYT, Vogel S. Global regulatory frameworks for the use of artificial intelligence (AI) in the healthcare services sector. Healthcare (Basel). 2024;12:562.
35. Adedinsewo DA, Pollak AW, Phillips SD, Smith TL, Svatikova A, Hayes SN, et al. Cardiovascular disease screening in women: leveraging artificial intelligence and digital tools. Circ Res. 2022;130:673–90.
36. Straw I, Wu H. Investigating for bias in healthcare algorithms: a sex-stratified analysis of supervised machine learning models in liver disease prediction. BMJ Health Care Inform. 2022;29:e100457.
37. Kormilitzin A, Tomasev N, McKee KR, Joyce DW. A participatory initiative to include LGBT+ voices in AI for mental health. Nat Med. 2023;29:10–1.
38. Cusick MM, Chertow GM, Owens DK, Williams MY, Rose S. The complexities of race adjustment in health algorithms. Stanford University Human-Centered Artificial Intelligence; 2024.

39. Abdulredah AA, Fadhel MA, Alzubaidi L, Duan Y, Kherallah M, Charfi F. Towards unbiased skin cancer classification using deep feature fusion. BMC Med Inform Decis Mak. 2025;25:48.
40. Perets O, Stagno E, Yehuda EB, McNichol M, Anthony Celi L, Rappoport N, et al. Inherent bias in electronic health records: a scoping review of sources of bias. medRxiv 2024. https://doi.org/10.1101/2024.04.09.24305594.
41. Chinta SV, Wang Z, Palikhe A, Zhang X, Kashif A, Smith MA, et al. AI-driven healthcare: fairness in AI healthcare: a survey. PLOS Digit Health. 2025;4:e0000864.
42. Obermeyer Z, Powers B, Vogeli C, Mullainathan S. Dissecting racial bias in an algorithm used to manage the health of populations. Science. 2019;366:447–53.
43. Bauer GR, Lizotte DJ. Artificial intelligence, intersectionality, and the future of public health. Am J Public Health. 2021;111:98–100.
44. Ncube M. Incomplete chronicles: unveiling data bias in maternal health; 2024.
45. Voinea C, Wangmo T, Vica C. Paternalistic AI: the case of aged care. Humanit Soc Sci Commun. 2024;11:824.
46. Tannenbaum C, Ellis RP, Eyssel F, Zou J, Schiebinger L. Sex and gender analysis improves science and engineering. Nature. 2019;575:137–46.
47. Achtari M, Salihu A, Muller O, Abbe E, Clair C, Schwarz J, et al. Gender bias in AI's perception of cardiovascular risk. J Med Internet Res. 2024;26:e54242.
48. Gendered Innovations. Heart disease in diverse populations: analyzing sex and gender. https://genderedinnovations.stanford.edu/case-studies/heart.html#tabs-2. Accessed 30 June 2025.
49. Trendall S. Gender bias concerns raised over GP app. 2019. Public Technology. https://www.publictechnology.net/2019/09/13/health-and-social-care/gender-bias-concerns-raised-over-gp-app/. Accessed 30 Jun 2025.
50. Li YH, Li YL, Wei MY, Li GY. Innovation and challenges of artificial intelligence technology in personalized healthcare. Sci Rep. 2024;14:18994.
51. De Martin TV, Wiig-Fisketjon A, Botten E, Dalen H, Langaas M, Bye A. Sex-specific cardiovascular disease risk prediction using statistical learning and explainable artificial intelligence: the HUNT Study. Eur J Prev Cardiol. 2025. https://doi.org/10.1093/eurjpc/zwaf135.
52. Sau A, Sieliwonczyk E, Patlatzoglou K, Pastika L, Mcgurk KA, Ribeiro AH, et al. Artificial intelligence-enhanced electrocardiography for the identification of a sex-related cardiovascular risk continuum: a retrospective cohort study. Lancet Digit Health. 2025;7:e184–94.
53. Rey S. AI model reads ECGs to identify female patients at higher risk of heart disease. Imperial's News site. 2025. https://www.imperial.ac.uk/news/261516/ai-model-reads-ecgs-identify-female/. Accessed 30 June 2025.
54. Ryu JS, Lee S, Chu Y, Ahn MS, Park YJ, Yang S. CoAt-Mixer: self-attention deep learning framework for left ventricular hypertrophy using electrocardiography. PLoS One. 2023;18:e0286916.
55. Ansari MY, Qaraqe M, Charafeddine F, Serpedin E, Righetti R, Qaraqe K. Estimating age and gender from electrocardiogram signals: a comprehensive review of the past decade. Artif Intell Med. 2023;146:102690.
56. Girlanda F, Demler O, Menze BH, Davoudi N. Enhancing cardiovascular disease prediction through multi-modal self-supervised learning. ArXiv. 2024;abs/2411.05900.
57. Dervishi A. A multimodal stacked ensemble model for cardiac output prediction utilizing cardiorespiratory interactions during general anesthesia. Sci Rep. 2024;14:7478.
58. Wenzl FA, Kraler S, Ambler G, Weston C, Herzog SA, Räber L, et al. Sex-specific evaluation and redevelopment of the GRACE score in non-ST-segment elevation acute coronary syndromes in populations from the UK and Switzerland: a multinational analysis with external cohort validation. Lancet. 2022;400:744–56.
59. Gendered Innovations. Methods of sex, gender, and intersectional analysis. https://genderedinnovations.stanford.edu/methods-sex-and-gender-analysis.html. Accessed 30 June 2025.
60. Tsao CW, Aday AW, Almarzooq ZI, Alonso A, Beaton AZ, Bittencourt MS, et al. Heart disease and stroke statistics-2022 update: a report from the American heart association. Circulation. 2022;145:E153–639.

61. Iribarren C, Chandra M, Lee C, Sanchez G, Sam DL, Azamian FF, et al. Breast arterial calcification: a novel cardiovascular risk enhancer among postmenopausal women. Circ Cardiovasc Imaging. 2022;15:e013526.
62. Wu H, Zhu YD, Shi WQ, Tong L, Wang MD. Fairness artificial intelligence in clinical decision support: mitigating effect of health disparity. IEEE J Biomed Health. 2025;29:815–23.
63. Kecki V, Said A, editors. Understanding fairness in recommender systems: a healthcare perspective. In: The 18th ACM conference on recommender systems (RecSys '24); 2024; Bari, Italy. New York, NY, USA: ACM.
64. Weerts H, Dudík M, Edgar R, Jalali A, Lutz R, Madaio M. Fairlearn: Assessing and improving fairness of AI systems. JMLR. 2023;24:1–8.
65. Luo Y, Khan MO, Wen C, Afzal MM, Wuermeling TF, Shi M, et al. FairDiffusion: enhancing equity in latent diffusion models via fair Bayesian perturbation. Sci Adv. 2025;11:eads4593.
66. Salih AMA, Raisi-Estabragh Z, Galazzo IB, Radeva P, Petersen SE, Lekadir K, et al. A perspective on explainable artificial intelligence methods: SHAP and LIME. Adv Intell Syst. 2023;7:2400304.
67. Gadgil S, DeGrave AJ, Daneshjou R, Lee S-I. Discovering mechanisms underlying medical AI prediction of protected attributes. medRxiv. 2024:2024.04.09.24305289.
68. Banerjee S, Alsop P, Jones L, Cardinal RN. Patient and public involvement to build trust in artificial intelligence: a framework, tools, and case studies. Patterns. 2022;3:100506.
69. Panigutti C, Perotti A, Panisson A, Bajardi P, Pedreschi D. FairLens: auditing black-box clinical decision support systems. Inf Process Manage. 2021;58:102657.
70. Saleiro P, Kuester B, Stevens A, Anisfeld A, Hinkson L, London J, et al. Aequitas: a bias and fairness audit toolkit. ArXiv. 2018;abs/1811.05577.
71. Kim C, Gadgil SU, Degrave AJ, Omiye JA, Cai ZR, Daneshjou R, et al. Transparent medical image AI via an image-text foundation model grounded in medical literature. Nat Med. 2024;30:1154–65.
72. Bellamy RKE, Dey K, Hind M, Hoffman SC, Houde S, Kannan K, et al. AI Fairness 360: an extensible toolkit for detecting and mitigating algorithmic bias. IBM J Res Dev. 2019;63:4:1–4:15.
73. Bragazzi NL, Garbarino S. Toward clinical generative AI: conceptual framework. JMIR AI. 2024;3:e55957.
74. Byrnes J, Robinson M. Transparency and authority concerns with using AI to make ethical recommendations in clinical settings. Nurs Ethics. 2024:9697330241307317.
75. European Commission. Artificial intelligence in healthcare. https://health.ec.europa.eu/ehealth-digital-health-and-care/artificial-intelligence-healthcare_en. Accessed 30 Jun 2025.
76. Health Data, Technology, and Interoperability: Certification Program Updates, Algorithm Transparency, and Information Sharing, 45 CFR Parts 170 and 171 RIN 0955–AA03. 2024.
77. Arbelaez Ossa L, Rost M, Bont N, Lorenzini G, Shaw D, Elger BS. Exploring patient participation in AI-supported health care: qualitative study. JMIR AI. 2025;4:e50781.
78. World Economic Forum. The future of AI-enabled health: leading the way. Geneva: Switzerland; 2025.
79. Pierce RL, Van Biesen W, Van Cauwenberge D, Decruyenaere J, Sterckx S. Explainability in medicine in an era of AI-based clinical decision support systems. Front Genet. 2022;13:903600.
80. Medical Science Sex and Gender Equity. Advancing sex and gender equity in UK biomedical, health and care research through policy co-design; 2025. https://www.messageproject.co.uk/. Accessed 30 Jun 2025.
81. Tan MJT, Benos PV. Addressing intersectionality, explainability, and ethics in AI-driven diagnostics: a rebuttal and call for transdisciplinary action. ArXiv. 2025;abs/2501.08497.
82. Liu M, Ning Y, Teixayavong S, Liu X, Mertens M, Shang Y, et al. A scoping review and evidence gap analysis of clinical AI fairness. NPJ Digit Med. 2025;8:360.